Fourth Edition

EARTH SCIENCE MANUAL

DEPARTMENT OF
EARTH SCIENCES & AG.

JAMES D. STEWART, CHAIRMAN

VINCENNES UNIVERSITY
VINCENNES, INDIANA

BURGESS INTERNATIONAL GROUP INC
Bellwether Press Division

Address orders to:

BURGESS INTERNATIONAL GROUP, Inc.
7110 Ohms Lane
Edina, Minnesota 55439-2143
Telephone 612/831-1344
EasyLink 629-106-44
Fax 612/831-3167

Bellwether Press
A Division of BURGESS INTERNATIONAL GROUP, Inc.

PREFACE

This manual of SES 100 Earth Science exercises is prepared exclusively for the SES 100 students of Vincennes University. It is the goal of the Earth Science faculty to provide our students with both a meaningful and enjoyable introductory course in the field of Earth Sciences.

The exercises in this manual cover a range of topics from Latitude and Longitude, Earth-Sun Relationships, and Mapping, to the Identification of Rocks and Minerals. For those students who plan to take additional Earth Science classes, this course will provide the necessary foundation upon which to expand; for those students who plan no further studies in Earth Science, this course will offer a well rounded exposure to the daily environmental events with which we live and should strive to better understand.

I wish to acknowledge the on-going input of the V.U. Earth Science faculty. The entire manual is the out-growth of years of combined teaching experience and topic selection by the Department faculty. Special thanks is hereby extended once again to Mr. Johnny Hart & Field Enterprises for their gracious consent in allowing the use of the B.C. Cartoons. An additional thanks is extended to Lissa Crowley for her excellent assistance in preparation of the manuscript.

It is with inspiration that I dedicate this manual to all Teachers and Students of Earth Science in the hope that they will find the interest and rewards that the study of Earth Science has provided me.

JAMES D. STEWART

CHAIRMAN, DEPARTMENT OF EARTH SCIENCE - AG
PROFESSOR OF EARTH SCIENCES
VINCENNES UNIVERSITY

TABLE OF CONTENTS

LABORATORY READING ASSIGNMENTS

EXERCISE	TEXT ASSIGNMENTS
#1	567 - 569
#2	337-341; 466-470
#3	567 - 569
#4*	NONE
#5	570 - 574
#6	570 - 574
#7	NONE
#8*	10 - 17; 564 - 566
#9	10 - 17; 564 - 566
#10	10 - 17; 564 - 566
#11*	17 - 18
#12	19 - 24
#13	24 - 35
#14*	ROCK EXAM ONLY

* LAB. EXAMINATIONS *
(1-3) (4-7) (8-10) (11-13)

SES 100   **COURSE DATA SHEET**   **EARTH SCIENCE**

- -

    Lecture Professor: _____ Sec. # _____

    Lecture Times: _____ in Room _____

    Laboratory Professor: _____ Sec. # _____

    Laboratory Times: _____ in Room _____

- -

Earth Science Faculty:

James D. Stewart, Chairman, Dept. of Earth Science & Ag
Professor of Earth Sciences
Office: MSC A-6
Office Telephone: 885-4235
Weekly Office Hours: _____

                              _____

William F. Bandy
Assistant Professor of Earth Sciences
Office: MSC B-1
Office Telephone: 885-4520
Weekly Office Hours: _____

                              _____

Neal E. Catt
Professor of Earth Sciences
Office: MSC B-3
Office Telephone: 885-4522
Weekly Office Hours: _____

                              _____

John A. Parsons
Professor of Earth Sciences
Office: MSC B-2
Office Telephone: 885-4521
Weekly Office Hours: _____

                              _____

Roy Wachter
Instructor of Earth Scienes
Office Hours by Appointment
- -

_____ Lab. Exercise 1 (10 pts)	_____ Lecture Quiz 1 (10 pts)
_____ Lab. Exercise 2 (10 pts)	_____ Lecture Exam 1 (100 pts)
_____ Lab. Exercise 3 (10 pts)	_____ Lecture Quiz 2 (10 pts)
_____ Lab. Exam I (50 pts)	_____ Lecture Exam 2 (100 pts)
_____ Lab. Exercise 4 (10 pts)	_____ Lecture Quiz 3 (10 pts)
_____ Lab. Exercise 5 (10 pts)	_____ Lecture Exam 3 (100 pts)
_____ Lab. Exercise 6 (10 pts)	_____ Lecture Quiz 4 (10 pts)
_____ Lab. Exercise 7 (10 pts)	_____ Lecture Exam 4 (100 pts)
_____ Lab. Exam II (50 pts)	
_____ Lab. Exercise 8 (10 pts)	_____ All Bonus Points
_____ Lab. Exercise 9 (10 pts)	
_____ Lab. Exercise 10 (10 pts)	_____ Final Exam (100 pts)
_____ Lab. Exam III (50 pts)	
_____ Lab. Exercise 11 (10 pts)	_____ Total Course Points (870)
_____ Lab. Exercise 12 (10 pts)	
_____ Lab. Exercise 13 (10 pts)	_____ Course %
_____ Lab. Exam IV (50 pts)	_____ Course Grade

Total pts. pos. / your total pts. = %

EXERCISE #1
LATITUDE AND LONGITUDE

A fundamental skill everyone should possess is the ability to pinpoint locations on the earth using globes and maps. Some caution must be used in the reading of globes and maps however since they are merely "representations" or "models" of the real world! Furthermore, the globe is a better representation of the earth because of its shape; all flat maps contain some error or distortion since they are 2 dimensional and are trying to represent a world which is 3 dimensional. Compare the size and shape of Greenland on a map with the true size and shape as shown on a globe.

Being able to correctly determine locations on the earth's surface is of great assistance to people in nearly all vocations. In order to pinpoint locations a universal coordinate system of direction is needed. The most common method used to determine such locations is that of latitude and longitude. Since the earth is spherical in shape it can be divided into equal units of 360°. Compass directions use the division of circles into 360° (bearings) for navigation purposes. In this system due North is 0° or 360°, East is 90°, South is 180°, and West is 270°.

Each of the 360° can further be divided into 60 equal units called "minutes" (NOTE: 1° = 60'), and each minute can also be subdivided into equal units called "seconds" (NOTE: 1' = 60"). Thus 10 degrees and 15 minutes of either latitude or longitude would be written 10° 15' 00". Another example is the writing of 2 minutes and 24 seconds; it would appear as follows: 00° 02' 24". Remember that latitude and longitude represents measurements of an arc; in other words, the angular division of a circle between two locations on the earth's surface!

The latitude and longitude coordinate system utilizes the equator to divide the earth into the northern and southern hemispheres; the Prime Meridian together with the International Date Line divides the earth into an eastern and western hemisphere. Latitude measures from 0° at the equator to 90° north/south (the location of the geographic or "true" North and South Poles). Longitude measures from 0° at the Prime Meridian to 180° east/west (the location of the International Date Line). These locations or coordinates together divide the earth into four quadrants. See <u>Figure 1.</u>

Latitude and longitude are thus always expressed in degrees, minutes, and seconds; however, it is not always possible to determine all three divisions because of the size of map area shown (scale of map). When it is impossible to determine the exact degrees, minutes, and/or seconds, merely insert zeros (00) to indicate the readings were not forgotten!

<u>IMPORTANT</u> <u>POINTS</u> <u>IN</u> <u>USING</u> <u>LATITUDE</u> & <u>LONGITUDE</u>:

1. Latitude readings are always listed first.

2. 90° N/S are the largest possible latitude readings.

3. 179° 59' 59" is the largest possible longitude reading that can be either East or West.

4. The largest possible minute or second reading is 59; never enter a reading of 60' or 60"; each should be rounded off to 1 of the next higher unit!

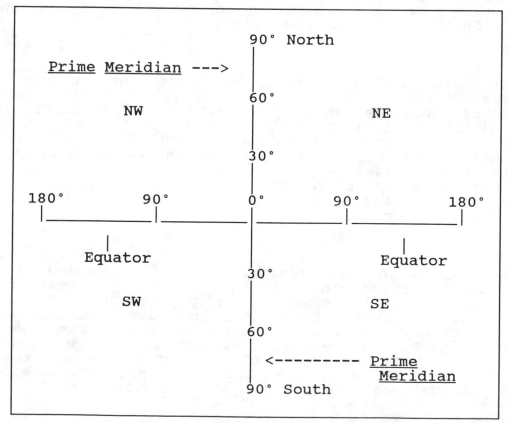

<u>Figure 1</u>

PRACTICE EXERCISES

Correct the following latitude and longitude readings:

1. 44° 10' 62" N   _45° 11' 2" N_
       45° 11' 2"
2. 85° 61' 50" S   _86° 60' 50_
       86° 60' 50
3. 155° 72' 62" E   _156° 74° 60" E_
4. 179° 59' 60" W   _179° 59' 60"_

INTERPOLATION OF LATITUDE AND LONGITUDE

Latitude: 22°25'

_____A_____ 22°   _____C_____ 6°

_____ 23°   _____ 5°

_____B_____ 24°   _____D_____ 4°

     23° 30' sec.

Answers: A = _22° 0' 0' South_   C = _5° 50' 0 N._
           B = _23° 30' 0' South_   D = _4° 15' 0' N_

Longitude:

   | | | | |
     A   B   C   D

40°   42°   44°   85°   83°   81°

                                 15
                              83° 0' 0' W

Answers: A = _41° 0' 0" E_   C = ~~83° 0' 0' W~~
           B = _42° 30° 0' E_   D = _80° 45' 45 W_

LATITUDE AND LONGITUDE PRACTICE EXERCISE

DETERMINE THE LATITUDE AND LONGITUDE FOR EACH OF THE FIVE
LOCATIONS IN THE "EARTH GRID" OF FIGURE 2.

- -

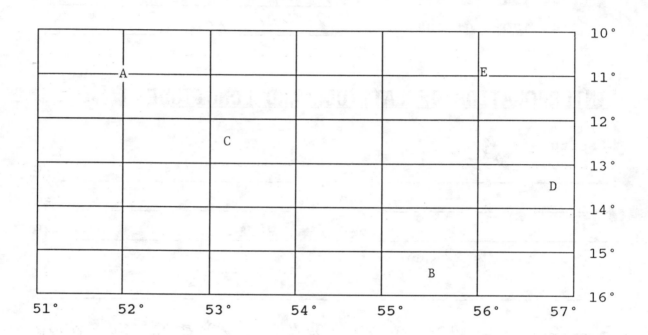

FIGURE 2

- -

ANSWERS

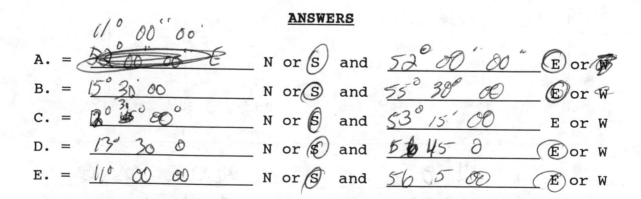

A. = 11° 00" 00'   N or (S) and 52° 00' 80" (E) or W

B. = 15° 30 00   N or (S) and 55° 30° 00 (E) or W

C. = 2° 45° 80°   N or (S) and 53° 15' 00 E or W

D. = 13° 30 0   N or (S) and 56 45 0   (E) or W

E. = 11° 00 00   N or (S) and 56 5 00   (E) or W

6

EARTH GRID

By establishing a system of N-S and E-W perpendicular lines across the earth's surface, we can produce a network referred to as a geographic or **"earth grid.** The E-W running lines are called "parallels" and the N-S running lines are called "meridians". Every point on the earth's surface can be located on this grid and no two places of different location can be described as having identical latitude and longitude coordinates!

By definition:

LATITUDE - "the distance in degrees N or S of the equator as determined by a series of lines (parallels) running E-W around the earth".

LONGITUDE - "the distance in degrees E or W of the Prime Meridian as determined by a series of lines (meridians) running N-S around the earth.

DISTANCE MEASUREMENT USING LATITUDE READINGS

Distances between locations can be determined by using latitude readings. $01°$ of latitude always equals 70 miles (rounded off). If two locations have the same longitude, you can easily figure the distance in miles they are apart by multiplying the degrees in latitude that separates them by 70 miles. **NOTE: this will not apply when locations have different longitudes!**

Further since there are 60' in $01°$ of latitude, each 01' is equal to 1/60th. of $01°$ (which is 1/60th. of 70 miles); therefore, 01' of latitude is equal to 1.17 miles or 6125 feet.

There is an identical calculation for 01" of latitude. Since there are 60" in 01' of latitude, 01" is equal to 1/60th of 01' (which is 1/60th. of 1.17 miles or 6125 feet); therefore, 01" of latitude is equal to 102 feet.

FOR REFERENCE:

1 Degree of Latitude = 70 miles

1 Minute of Latitude = 1.17 miles or 6125 feet

1 Second of Latitude = 102 feet

NOTE: The value of 01° of longitude, unlike latitude, is not constant! A degree of longitude varies from a value of 70 miles at the equator to 0 miles at the poles! This can easily be seen by observing any globe and noticing how parallels are always the same distance apart, regardless of location, while all meridians converge (become closer together) as they move toward the poles.
At 90° N or S there is no longitude!

PRACTICE EXERCISES

1. Quebec, Canada is located at 45° N. and is on the same meridian as Maracibo, Venezuela at 15° N. Approximately how many miles separate these two cities?

2. Lima, Peru is located at 12° S. and is on the same meridian as Washington, D.C. at 38° N. Approximately how many miles separate these two cities?

3. Vincennes, Indiana is located at 38° N. Approximately how many miles is Vincennes from the North Pole?

4. Location x is located at 12° 45' 39" N and 112° 16' 44" E. Location y is located at 39° 12' 20" S and 148° 56' 22" W. Can you determine how many miles separate x and y? If not, why?

5. Using what you have learned about latitude and longitude, work this problem: you are located at 15° N. and 123° E. If you moved 28° south, and 140° west, what would be your new latitude and longitude?

8

SIGNIFICANT PARALLELS

Equator .... Divides the earth into a N & S Hemisphere; located at 0° N/S

Tropic of Cancer .... Northern most latitude that the sun's vertical rays ever reach; Located at 23 $1/2^\circ$ N.

Tropic of Capricorn ...... Southern most latitude that the sun's vertical rays ever reach; Located at 23 $1/2^\circ$ S.

Arctic Circle .... Northern most latitude that the sun's rays reach during winter. Located at 66 $1/2^\circ$ N.

Antarctic Circle .... Southern most latitude that the sun's rays reach during summer. Located at 66 $1/2^\circ$ S.

SIGNIFICANT MERIDIANS

Prime Meridian .... Passes through Greenwich, England & represents 0° E/W. Creates E & W Hemisphere.

International Date Line .. Lies directly opposite the Prime Meridian & represents 180° E/W. Each day starts and ends here!

GREAT CIRCLE DISCUSSION

By definition, this is a circle drawn around the earth whose plane passes through the earth's center thereby cutting it into 2 equal hemispheres.

You should see that all opposing meridians form great circles, but the only parallel which is a great circle is the equator! Any two locations on earth are connected by a great circle which passes through them and cuts the earth into equal halves. The shortest distance between any two locations on earth follows the great circle route connecting them. This fact is illustrated by the routes selected by ships and airplanes during international travel. Your laboratory instructor will demonstrate these routes.

PRACTICE EXERCISES

On the following 3 maps (Continents, United States, & Indiana), you are to determine the latitude and longitue for the locations listed below and shown on the maps by corresponding letters. For the Continents and U.S. maps, you are given the latitude and longitude coordinates and will place letters for the locations on the maps! Using the Indiana map, you are to determine the location of the 2 listed cities to the **nearest minute!**

Notice how the parallels and meridians curve on the maps! Be sure to follow the lines as they bend!

Locate the following on the Continents Map:

A. Lake Balkhash .... 46° N. and 75° E.

B. Vancouver Island .... 49° N. and 125° W.

Locate the following on the United States Map:

C. Lake Okeechobee .... 27° N. and 80° 49' W.

D. New Orleans .... 30° N. and 90° 05' W.

Determine the Latitude and Longitude for the following Indiana cities located on the Indiana Map (to the nearest minute)!

E. Vincennes .... Latitude = _____

                              Longitude = _____

F. Fort Wayne .... Latitude = _____

                              Longitude = _____

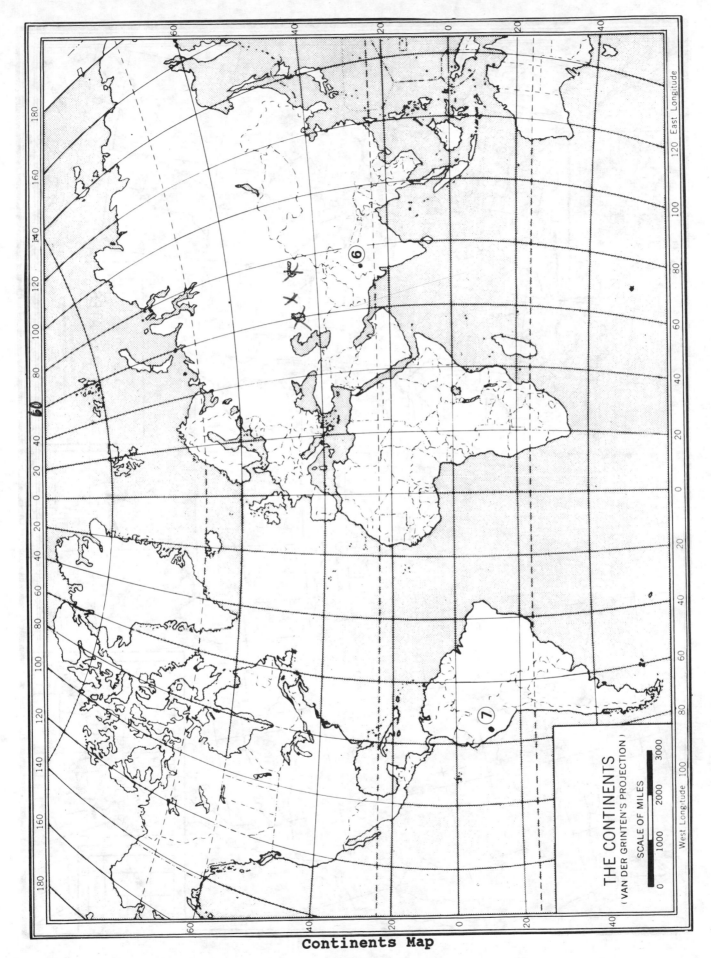

Continents Map

11

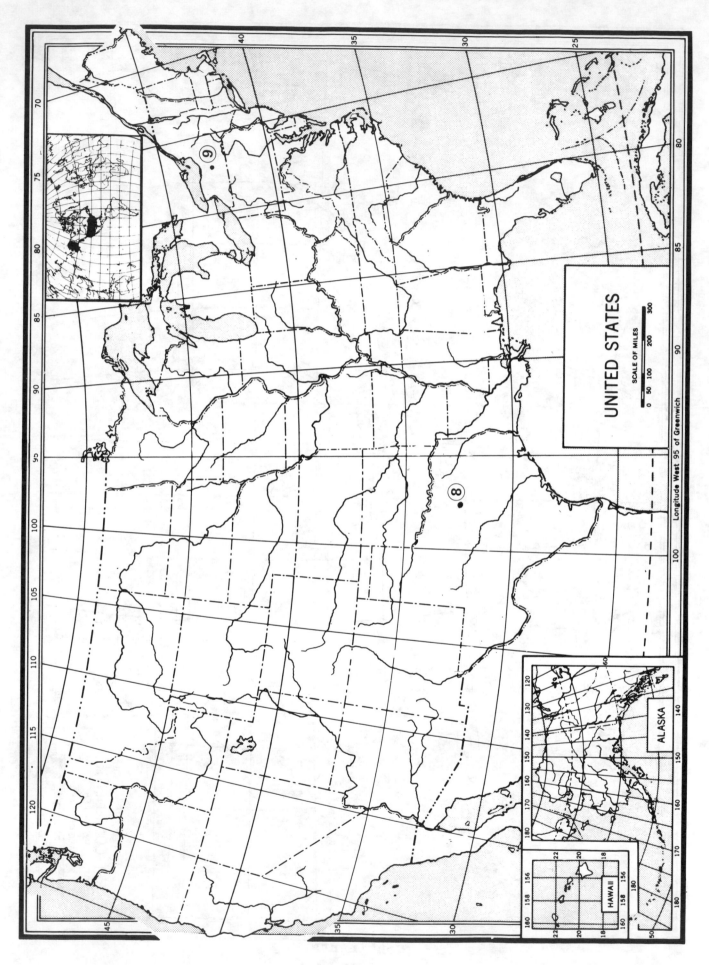

UNITED STATES

SCALE OF MILES

0 50 100   200   300

Longitude West 95 of Greenwich

ALASKA

HAWAII

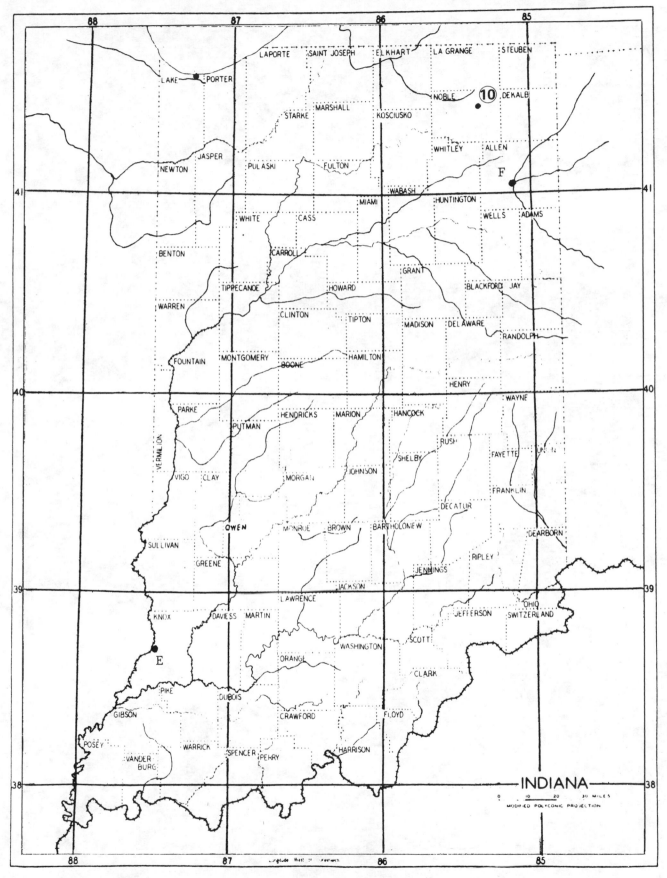

Indiana Map

13

EXERCISE 1-A
LATITUDE AND LONGITUDE

Student Name: _____   Lab. Sec. # _____

1. Latitude is determined by a series of
    lines running north and south across
    the earth's surface; these lines are
    known as meridians.   TRUE   FALSE

2. It is possible for a location on earth
    to have a latitude reading of 90° 15' 58"
    South.   TRUE   FALSE

3. Walking from 15° East to 30° East is a
    greater distance at the equator than at
    the 45° South parallel.   TRUE   FALSE

4. There is no such place on earth as 00°
    latitude and 00° longitude.   TRUE   FALSE

5. You are located at 75° North, and 102° West.
    If you move 100° S, and 100° E, what
    is your location?   _____

**USING AN ATLAS PROVIDED, DETERMINE THE LATITUDE & LONGITUDE
FOR THE FOLLOWING CITIES TO THE NEAREST DEGREE** (Minutes &
Seconds Cannot Be Determined)! **NOTE: Page numbers for the
locations will be provided by your instructor!**

6. Fairbanks, Alaska ...... _____

7. Sao Paulo, Brazil ...... _____

8. Berlin, East Germany ... _____

9. Addis Ababa, Ethiopia .. _____

10. Manila, Philippines .... _____

EXERCISE 1-B
LATITUDE AND LONGITUDE

Student Name: _____ Lab. Sec. # _____

1. While most great circles appear curved
    on maps, it is possible for some to appear
    as straight lines.   TRUE   FALSE

2. As one moves farther away from the earth's
    poles, a degree of longitude represents
    fewer miles.   TRUE   FALSE

3. A latitude reading of 06° 14' 22" is the
    same as a reading of 05° 74' 22".   TRUE   FALSE

4. If you were located at 165° East and moved
    20° farther to the east, your new location
    would be 175° West.   TRUE   FALSE

5. You could get more accurate latitude and
    longitude readings from a map of a state
    than of a continent.   TRUE   FALSE

**USING AN ATLAS PROVIDED, DETERMINE THE LATITUDE & LONGITUDE
FOR THE FOLLOWING CITIES TO THE NEAREST DEGREE (Minutes &
Seconds Cannot Be Determined)! NOTE: Page numbers for the
locations will be provided by your instructor!**

6. Lagos, Nigeria .... _____

7. Ulan Bator, Mongolia .... _____

8. Athens, Greece .... _____

9. Madrid, Spain .... _____

10. Montevideo, Uruguay ..... _____

EXERCISE 1-C
LATITUDE AND LONGITUDE

Student Name: _____ Lab Sec. # _____

1. The accuracy of latitude and longitude readings is influenced by the size of the area covered by a map.   TRUE   FALSE

2. The longitude reading 165° 16' 95" is incorrect; correctly written it reads 165° 17' 35".   TRUE   FALSE

3. If you walk 1 mile or (5280') due north or south across the earth's surface you have moved by over 01' of latitude.   TRUE FALSE

4. As you cross the Prime Meridian while traveling west you are entering the Western Hemisphere.   TRUE   FALSE

5. You are located at 30° North and 40° West. If you move 42° south, and 50° east, what is your location?   _____

USING AN ATLAS PROVIDED, DETERMINE THE LATITUDE & LONGITUDE FOR THE FOLLOWING CITIES TO THE NEAREST DEGREE (Minutes & Seconds Cannot Be Determined)! **NOTE: Page numbers for the locations will be provided by your instructor!**

6. Hobart, Tasmania .... _____

7. Lusaka, Zambia .... _____

8. Paris, France .... _____

9. Santiago, Chile .... _____

10. Regina, Canada .... _____

EXERCISE 1-D
LATITUDE AND LONGITUDE

Student Name: _____ Lab. Sec. # _____

1. The Prime Meridian together with the
    International Date Line forms a
    "great circle".   TRUE   FALSE

2. The latitude reading 15° 75' 61" is
    incorrect; correctly written it reads
    16° 15' 01".   TRUE   FALSE

3. Two locations on the same meridian are
    separted by 30° of latitude; thus, they
    are over 2000 miles apart.   TRUE   FALSE

4. It is approximately 50 miles from
    Vincennes to Evansville; in latitude
    that is over 10".   TRUE   FALSE

5. You are located at 54° South and 140°
    East. If you move 24° north, and 15°
    west, what is your location?   _____

**USING AN ATLAS PROVIDED, DETERMINE THE LATITUDE & LONGITUDE
FOR THE FOLLOWING CITIES TO THE NEAREST DEGREE** (Minutes &
Seconds Cannot Be Determined! **NOTE: Page numbers for the
locations will be provided by your instructor!**

6. Canberra, Australia .... _____

7. Colombo, Sri Lanka .... _____

8. Athens, Greece .... _____

9. Quito, Ecuador .... _____

10. Edmonton, Canada .... _____

EXERCISE 1-E
LATITUDE AND LONGITUDE

Student Name: _____ Lab. Sec. # ____

1. Degrees north or south of the equator
    represents the longitude of a location.   TRUE FALSE

2. Walking 01° of latitude across the earth's
    surface is the same distance regardless
    of the longitude.   TRUE FALSE

3. If you started at the equator and walked
    straight ahead, always toward a pole, for
    135°, you could end up at a latitude
    of 45°.   TRUE FALSE

4. If you moved 3/4 ths. of 01° of longitude
    to the west of the Prime Meridian you
    would be at 00° 45' 00" W.   TRUE FALSE

5. You are located at 48° North and 32° West.
    If you move 52° south, and 78° east,
    what is your location?   _____

**USING AN ATLAS PROVIDED, DETERMINE THE LATITUDE & LONGITUDE
FOR THE FOLLOWING CITIES TO THE NEAREST DEGREE (Minutes &
Seconds Cannot Be Determined)! NOTE: Page numbers for the
locations will be provided by your instructor!**

6. Tripoli, Libya .... _____

7. Moscow, Russia .... _____

8. Brisbane, Australia ..... _____

9. Bombay, India .... _____

10. Dallas, Texas .... _____

EXERCISE 1-F
LATITUDE AND LONGITUDE

Student Name: _____ Lab. Sec. # _____

1. The closer one gets to the equator, the farther apart meridians become in distance.   TRUE   FALSE

2. 01° 31' 15" of longitude is the same as 5424 minutes of longitude.   TRUE   FALSE

3. It is possible for a "great circle" route which passes through Vincennes to also be parallel to the equator.   TRUE   FALSE

4. A degree of longitude represents the same distance all over the earth; the same is not true for latitude.   TRUE   FALSE

5. You are located at 15° North and 11° East. If you move 65° south, and 70° west what is your location?   _____

USING AN ATLAS PROVIDED, DETERMINE THE LATITUDE & LONGITUDE FOR THE FOLLOWING CITIES TO THE NEAREST DEGREE (Minutes & Seconds Cannot Be Determined)! NOTE: Page numbers for the locations will be provided by your instructor!

6. Tokyo, Japan .... _____

7. Leningrad, Russia ..... _____

8. Oslo, Norway .... _____

9. LaPaz, Bolivia .... _____

10. Mexico City, Mexico ... _____

EARTH SCIENCE   LAB. EXERCISE #2

"EARTH-SUN RELATIONSHIPS"

OBJECTIVES:

BE ABLE TO:

1. EXPLAIN THE CAUSES OF:

   A. VARIATIONS IN LENGTH OF DAYLIGHT;
   B. SEASONAL CLIMATIC CHANGES;
   C. CHANGE IN DIRECTION OF SUNRISE/SUNSET

2. DISCUSS THE IMPORTANCE OF THE PLANE OF THE ECLIPTIC & CONSTANT PARALLELISM.

3. ASSOCIATE PERIHELION & APHELION WITH THE RESPECTIVE DATES AND DISTANCES.

4. USE DIAGRAMS TO SHOW THE RELATIONSHIP BETWEEN EARTH & SUN FOR EACH SEASON.

5. ASSOCIATE THE CORRECT NAME & DATE FOR THE FIRST DAY OF EACH OF THE SEASONS.

6. DISCUSS THE CIRCLE OF ILLUMINATION.

7. DETERMINE SUNRISE/SUNSET DIRECTIONS & LOCATION OF SUN'S VERTICAL RAYS.

8. USE THE ANALEMMA FOR ANY DATE.

9. DEFINE THE GLOSSARY TERMS.

B.C. by permission of Johnny Hart and Field Enterprises, Inc.

GLOSSARY TERMS

Rotation:

Revolution:

Aphelion:

Perhelion:

Plane of Ecliptic:

Inclination of Axis:

Constant Parallelism:

Equinox:

Solstice:

Arctic Circle:

Antarctic Circle:

Tropic of Cancer:

Tropic of Capricorn:

Equator:

Circle of Illumination:

Vertical Rays:

Oblique Rays:

Tangent Rays:

Analemma:

EXERCISE #2

EARTH-SUN RELATIONSHIPS

The earth is but one of nine planets revolving around a medium sized star known as the sun. This sun and all bodies confined to revolution around it constitutes our solar system. The solar system likewise rotates within our galaxy (the Milky Way), and the galaxy is drifting as an infinitesimal object within the vastness of the universe.

Of the many various motions which involve the earth and solar system, the most influential to man are rotation and revolution. In this exercise, consideration will be given to these two motions and their relationship to such phenomena as:

1. variation in hours of daylight and darkness;
2. seasonal climatic changes;
3. seasonal changes in direction/time of sunrise/sunet

There are four basic earth characteristics that are responsible for the phenomena mentioned above; these four basic earth traits (the basis of concern for this exercise) are:

1. earth rotation;
2. earth revolution;
3. inclination of earth's axis;
4. constant parallelism of earth's axis.

EARTH ROTATION

The fact that the earth rotates (turns) on an axis every 24 hrs. (actually 23 hours & 56 minutes), thus creating the periods of daylight and darkness, has become a natural clock for mankind. This movement of the earth is west to east (counterclockwise) as viewed from the North Star (Polaris), resulting in a sunrise in the East and sunset in the West. Earth rotation is known to be slowing down because of the tidal friction of ocean waters upon the earth's surface. The result of this slowing of rotation is a gradual increase in the length of the earth day at the rate of 1 additional second/day for each 1000 years!

29

EARTH REVOLUTION

Revolution refers to the movement of the earth around the sun in a counterclockwise direction. One complete revolution takes approximately 365 1/4 days -- the basis for our calendar year. Although not perfectly circular, deviation from a perfect orbit is only about 3 percent,. The earth reaches **perihelion (closest to sun)** in early to middle January, and **aphelion (farthest from sun)** in early to middle July. During aphelion, the earth and sun are separated by 94,000,000 miles; at perihelion the distance separating earth and sun is 91,500,000 miles.

The variation in distance from the sun actually results in minor differences in the amount of solar energy received on earth during revolution. It therefore is no the cause of seasonal climatic change. This fact is evident when you consider that the Northern Hemisphere receives its warmest weather during July at the time when the earth is farthest from the sun! **The most significant aspect of earth-sun relationships is the orientation of the earth toward the sun and the resulting differences in the intensity of solar rays striking the earth's surface! (See Figure 5)**

The 2 reasons for warmer July and colder January temperatures in the northern hemisphere are the angle at which the sun's rays hit the earth, and the length of daylight received.

1. **Angle of the Sun** -- this refers to the altitude of the sun above the horizon when the sun reaches its highest point in the sky (12:00 noon local time). Using Vincennes as an example, on June 21 the sun at noon is 75 degrees above our southern horizon. On December 22, the sun at noon is only 28 degrees above our southern horizon. Due to curvature of the earth, the same rays of sunlight will strike the earth more vertical in summer than winter and therefore provide more of a heating effect. There is also more surface in the northern hemisphere receiving solar rays in the summer than in the winter months.

2. **Length of Daylight** -- again using Vincennes, June 21 is our longest day of the year. On this date, Vincennes has nearly 15 hours of daylight and only 9 hours of darkness. On December 22, our shortest day of the year, the reverse is true (we experience 9 hours of daylight and 15 hours of darkness). Obviously, the longer the day the greater the amount of heating!

EARTH INCLINTION AND CONSTANT PARALLELISM

Throughout revolution the earth is confined to a plane known as the **ecliptic**. This plane is an imaginary surface which passes through both the center of the sun and earth. The earth is tilted on its axis 23 1/2 degrees in reference to the place of ecliptic. Since the earth maintains a position of **constant parallelism** at all times throught its orbit, the axis of the earth is always pointed toward the same point in space. Locations on earth are thus always tilted toward, away from, or parallel to the sun depending upon the date (determined by the location of the earth in its revolution around the sun).

It is therefore the earth's 23 1/2 degree axis inclination along with constant parallelism that is responsible for seasonal change. **Be sure to learn the dates for the first day of each of the four seasons, and the corresponding orientation of the earth axis in reference to the sun!**

AXIAL ORIENTATION OF EARTH ON SEASONAL DATES: See Figure 3.

June 21-22 . . . . North Pole toward Sun
Summer Solstice (Start of Summer)

September 21-22 . . . . .Poles Equidistant from Sun
Autumnal Equinox (Start of Fall)

December 21-22 . . . . South Pole toward Sun
Winter Solstice (Start of Winter)

March 21-22 . . . . .Poles Equidistant from Sun
Vernal Equinox (Start of Spring

EARTH POSITIONS WITH RESPECT TO THE SUN

The spherical shape of the earth is responsible for the changing angle of the sun and the change in length of day throughout the year. Due to this shape, the earth is always half in light and half in dark. The boundary line which separates the two spheres of light and dark around the globe has been given the name **"circle of illumination"**; its location changes throughout the year as a study of the seasonal diagrams will show. **Be sure to notice how the circle of illumination shifts back and forth as the seasons change (this is especially noticable and important at the poles!).**

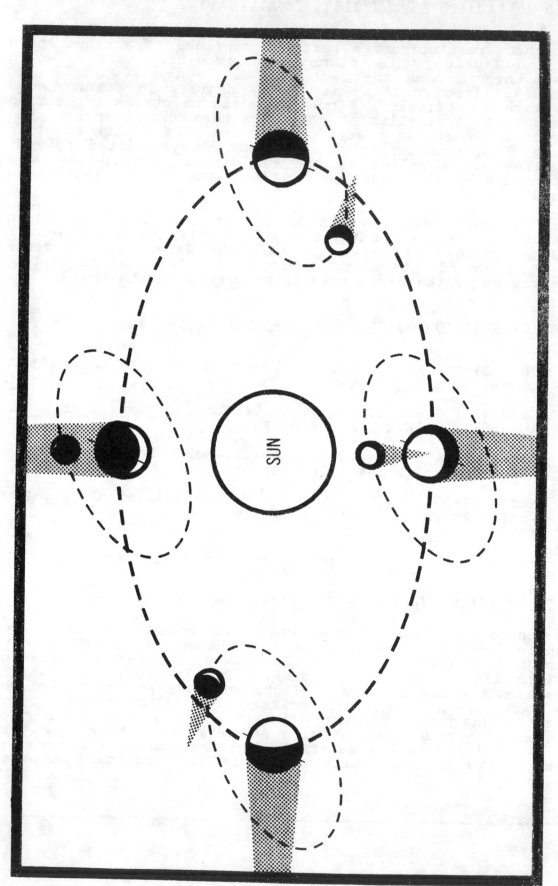

SEASONAL CHANGES

Figure 3.

ANGLE OF SUN'S RAYS STRIKING THE EARTH

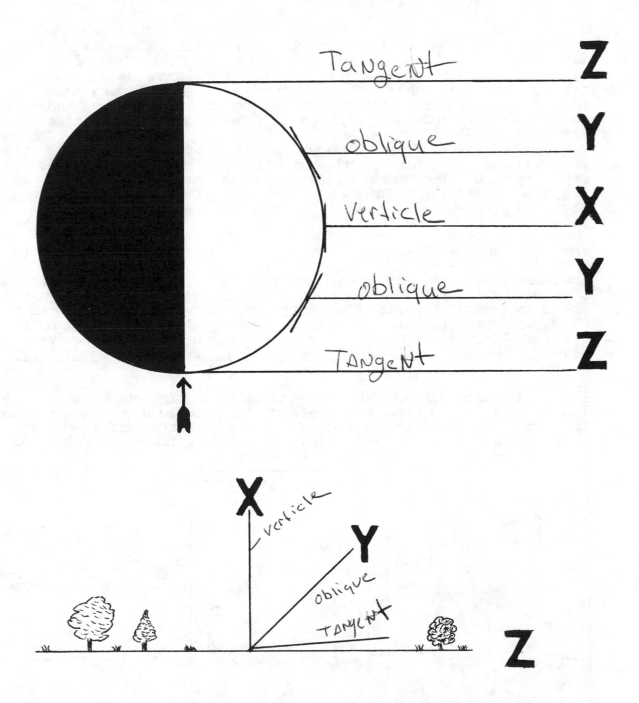

Figure 4

VERTICAL, TANGENT AND OBLIQUE RAYS

Since the sun is so much larger than the earth (108x larger) and so distant (average 93,000,000 miles), all rays of sunlight that strike the earth are parallel. Due to the spherical shape of the earth there are always vertical, tangent, and oblique rays of sunlight striking the earth's surface. (See diagram below).

Tangent rays pass the earth and mark the circle of illumination; the half of the earth behind the circle of illumination in reference to the sun is always dark. Oblique rays strike the earth from angles of 1 to 89 degrees; the larger the angle, the more heating that results and the longer the length of daylight! Vertical rays strike the earth at a 90 degree angle (directly overhead). These vertical rays strike a different latitude each day of the year moving as far south as 23 1/2 degrees (Tropic of Capricorn) on December 21-22, and as far north as 23 1/2 degrees (Tropic of Cancer) on June 21-22. (See Analemma: Figure 6)

The latitude of the sun's vertical rays for each day of the year is shown on the diagram of the analemma. On the analemma the day and month of the year are written on the figure eight, and the degrees north or south of the equator are shown on the left side of the graph. The analemma can be used as a navigation aid since for any day of the year the latitude for the location of the sun's vertical rays is given. Where are the sun's vertical rays on your birthday?

Any observer, by noting the altitude of the noon sun above the horizon and using the analemma to determine where the sun is vertical, cane asily calculate his latitudinal position. Such procedure will be explained outside the class by your lab. instructor if you are interested!

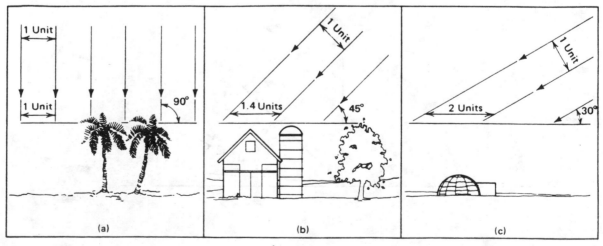

Figure 5

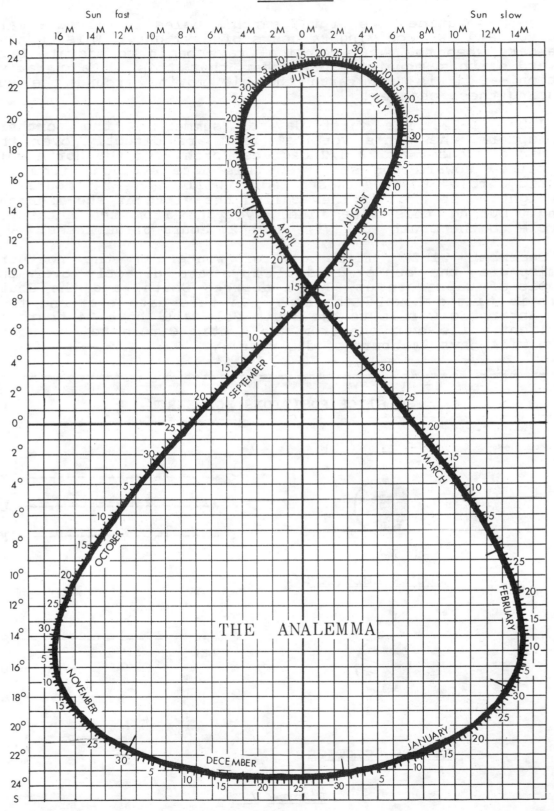

Figure 6

SOLSTICE:

During a solstice, one of the earth's poles is tilted toward the sun and is receiving 24 hours of daylight. This is also the date when the sun's vertical rays begin to migrate back toward earth's equator. Remember that this is a relative motion -- it is the earth that is moving around the sun causing these changes!

On June 21-22, the first day of summer in the northern hemisphere, the sun's vertical rays are directly over the Tropic of Cancer at 23 1/2 degrees North, and the North Pole is tilted toward the sun. On this date everyone from the Arctic Circle to the North Pole will have 24 hours of daylight! The area is then called the **"land of the midnight sun".** During this same time, everyone from the Antarctic Circle to the South Pole will have 24 hours of darkness!

On December 21-22, first day of winter in the northern hemisphere, the sun's vertical rays are directly over the Tropic of Capricorn at 23 1/2 degrees South, and the South Pole is tilted toward the sun. On this date everyone from the Antarctic Circle to the South Pole will have 24 hours of daylight while everyone from the Arctic Circle to the North Pole will have 24 hours of darkness!

SOLSTICE POSITIONS FOR EARTH

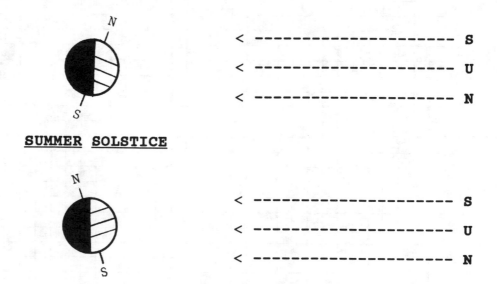

SUMMER SOLSTICE

WINTER SOLSTICE

Figure 7.

EQUINOX:

During an equinox both poles of earth are equidistant from the sun and the sun's vertical rays are over the equator. All locations on earth will receive 12 hours of both day and night on this date!

The Vernal Equinox occurs on March 21-22 and represents the first day of Spring in the northern hemisphere. The Autumnal Equinox occurs on September 21-22 and represents the first day of Autumn (Fall) in the northern hemisphere. **Remember that the seasons are reversed in the southern hemisphere!**

Days at the North and South Pole last for 6 months. On March 21-22 at 12:00 noon, the sun will rise above the horizon, but won't set again until noon on September 21-22. Equatorial locations, on the other hand, receive equal amounts of daylight and darkness, with very little variation, throughout the year regardless of season.

Another important facet of seasonal change for earth throughout a year is the reversal of seasons between the North and South Hemispheres. Opposite seasons always exist for the hemispheres due to the direction of axial tilt (either toward or away from the sun). The first day of summer in the Northern Hemisphere is the first day of winter in the Southern Hemisphere and so forth.

Note: Since the relationship between the earth and sun is the same for both equinox seasons, only one diagram is needed to show the equinox position relationships!

EQUINOX POSITION FOR EARTH

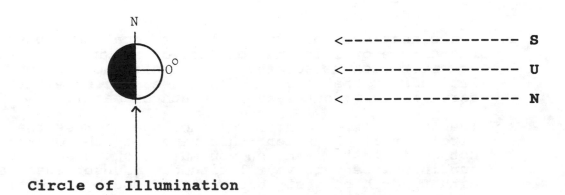

Circle of Illumination

MIGRATION OF SUN'S VERTICAL RAYS & SUNRISE AND SUNSET DIRECTIONS

You should have noticed the sun's vertical rays are found only within the tropics (between the Tropics of Cancer & Capricorn) regardless of the time of year! The noon sun therefore moves across a total of 47 degrees of latitude throughout the year.

The astronomical year begins on March 21-2 as the vertical rays from the sun move north of the equator. On June 21-22, vertical rays from the sun have reached the Tropic of Cancer (23 1/2 N.) and stop moving north on that date. The term **"solstice"** means **"sun stand still"**! After June 21-22, the vertical rays from the sun reverse movement and begin a southward migration toward the equator reaching that location on September 21-22. By December 21-22, the vertical rays of the sun have reached the Tropic of Capricorn (23 1/2 S.) and then stop their southward migration; a reversal of direction starts them back toward the equator again and the cycle is never ending!

MIGRATION OF SUN'S VERTICAL RAYS

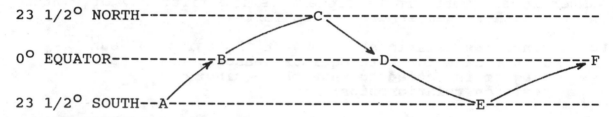

```
23 1/2° NORTH----------------------->C----------------------------------

0° EQUATOR------------->B------------------>D------------------------>F

23 1/2° SOUTH--A---------------------------------------E------------
```

WHAT SEASON FOR THE NORTHERN HEMISPHERE IS REPRESENTED BY EACH OF THE LETTERS (A-F) IN THE DIAGRAM ABOVE?

SUNRISE & SUNSET DIRECTIONS

1. You cannot see the sun until your location crosses the circle of illumination; notice in what direction you must look to see the sun in reference to your latitude as that occurs. (Sunrise must be either north of east, due east, or south of east).

2. The sun won't set until you once again cross the circle of illumination (this time into the dark); notice in what direction you must look to see the sun in referene to your latitude. (Sunset must be either north of west, due west, or south of west).

3. Location on earth is of no importance in determining the direction of sunrise/sunset--directions are same for all!

PRACTICE EXERCISES

Complete the drawing below to show the position of the earth with reference to the sun for December 21-22. Label all the important parallels, and mark the circle of illumination.

S ------------------->

U ------------------->

N ------------------->

December 21-22

How many hours of daylight would there by on December 21-22 at:

Equator __12__   72° S. __24__   68° N. __0__

What direction would people in Santiago, Chile look to see the sunrise on this date? __South of East__

What direction would people in Denver, Colorado look to see the sunset on this date? __Southwest__

S ------------------->

U ------------------->

N ------------------->

June 21-22

How many hours of daylight would there be on June 21-22 at:

Equator __12__   75° S. __0__   67° N. __24__

What direction would people in Perth, Australia look to see the sunrise on this date? __NE__

What direction would people in Paris, France look to see the sunset on this date? __NW__

Would earth climates be different if the axis was not tilted?

IN SUMMARY:

1. **VERNAL (Spring) EQUINOX:**   Date - March 21

    a. North and South Poles are equal distant from the sun.
    b. Equal hours (12/12) of day & night everywhere on earth.
    c. Sun's vertical rays strike earth at the equator.
    d. First day of spring for Northern Hemisphere.
    e. First day of fall for Southern Hemisphere.

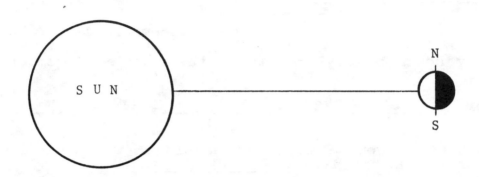

2. **SUMMER SOLSTICE:**   Date - June 21

    a. North Pole is oriented _toward_ the sun.
    b. Sun's vertical rays strike earth at 23 1/2° N.
    c. Area between 66 1/2° N and 90° N has 24 hours of daylight.
    d. Area between 66 1/2° S and 90° S has 24 hours of darkness.
    e. Northern Hemisphere has long days and short nights.
    f. Southern Hemisphere has short days & long nights.
    g. First day of summer for Northern Hemisphere.
    h. First day of winter for Southern Hemisphere.

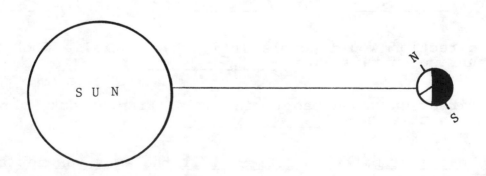

3. **AUTUMNAL (Fall) EQUINOX:**   Date - September 21

   a. <u>Same</u> as the Vernal Equinox (opposite side of the
      earth's orbit around the sun).
   b. First day of fall for Northern Hemisphere.
   c. First day of spring for Southern Hemisphere.

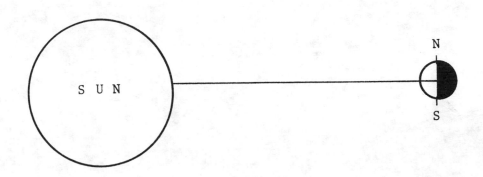

4. **WINTER SOLSTICE:**   Date - December 22

   a. North Pole is oriented <u>away</u> <u>from</u> the sun.
   b. Sun's vertical rays strike earth at 23 1/2° S.
   c. Area between 66 1/2° N and 90° N has 24 hours of
      darkness.
   d. Area between 66 1/2° S and 90° S has 24 hours of
      daylight.
   e. Northern Hemisphere has short days & long nights.
   f. Southern Hemisphere has long days & short nights.
   g. First day of winter for Northern Hemisphere.
   h. First day of summer for Southern Hemisphere.

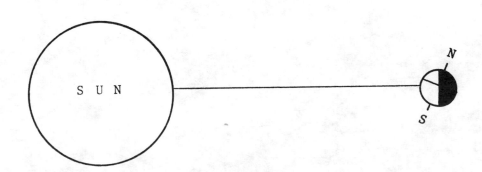

EXERCISE 2-A
EARTH-SUN RELATIONSHIPS

Student Name: _____   Lab. Sec. # _____

1. The vertical rays are within 4° of the
    equator on April 1.   TRUE   FALSE

2. During the summer solstice (June 21) the
    North Pole of earth is tilted away from
    the sun.   TRUE   FALSE

3. When the earth is at perihelion, the northern
    hemisphere receives the greatest amount of
    solar heat.   TRUE   FALSE

4. On June 21 a person located at 65° N.
    would not experience sunset or darkness
    throughout a 24 hr. interveral.   TRUE   FALSE

5. On August 20, everyone on earth could see
    the sunrise north of East: everyone could
    also see the sunset north of West later on
    the same date.   TRUE FALSE

6. - 10.

    The date is June 21. Using the earth diagram below:

    a. **draw and label** parallels **(use ruler);**

    b. **draw** the circle of illumination;

    c. what is the sunrise direction?   E   NE   SE

    d. what is the sunset direction?   W   NW   SW

    e. Southern Hemisphere season? _____

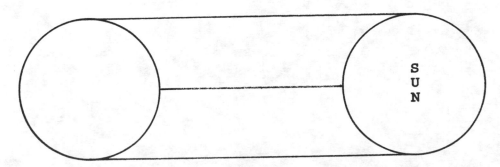

EXERCISE 2-B
EARTH-SUN RELATIONSHIPS

Student Name: _____   Lab. Sec. # _____

1. The noon sun will be directly overhead to
    all locations on earth at least once every
    year.   TRUE   FALSE

2. The sun's vertical rays are within 5°
    of the Tropic of Cancer on August 19
    every year.   TRUE   FALSE

3. Because of tidal friction, the earth's
    rotation is slowing down resulting in
    longer days.   TRUE   FALSE

4. The sun is not visible at 56° N. on
    June 21.   TRUE   FALSE

5. On December 22 each year, everyone on
    earth will see sunrise due east, and
    sunset due west.   TRUE   FALSE

6. - 10.

   The date is Sept. 21. Using the earth diagram below:

   a. **draw & label** parallels **(use ruler)**;

   b. **draw** the circle of illumination;

   c. what is the sunrise direction?   E   NE   SE

   d. what is the sunset direction?   W   NW   SW

   e. Northern Hemisphere season? _____

EXERCISE 2-C
EARTH-SUN RELATIONSHIPS

Student Name: _____   Lab. Sec. # _____

1. On March 4, the vertical rays from the sun
   strike earth's surface within 5° of the
   equator.   TRUE   FALSE

2. Oblique rays from the sun strike the earth's
   surface at the Tropic of Capricorn on
   December 22.   TRUE   FALSE

3. The longest day of the year occurs in
   the N. Hemisphere when the S. Hemisphere
   has its shortest day.   TRUE   FALSE

4. The changing distance between the earth
   and sun is the major cause for change
   of earth seasons.   TRUE   FALSE

5. People in both N & S Hemispheres will see
   a northeast sunrise during the months of
   June & July.   TRUE   FALSE

6. - 10.

   The date is Dec. 21. Using the earth diagram below:

   a. **draw & label** parallels **(use ruler)**;

   b. **draw** the circle of illumination;

   c. what is the sunrise direction?   E   NE   SE

   d. what is the sunset direction?   W   NW   SW

   e. Northern Hemisphere season? _____

EXERCISE 2-D
EARTH-SUN RELATIONSHIPS

Student Name: _____   Lab. Sec. # _____

1. At what latitude do the vertical rays
    from the sun strike earth on
    November 25? **(Use Anelemma)**   _____

2. Regardless of the date, people in Indiana
    never see the sun directly overhead.   TRUE   FALSE

3. If the earth's axis was perpendicular to
    the plane of ecliptic instead of inclined,
    there would be less seasonal change.   TRUE   FALSE

4. On January 15, everyone on earth sees a SE
    sunrise and a SW sunset.   TRUE   FALSE

5. On September 21, the sun cannot be seen
    any hour of the day at the latitude of
    86° N.   TRUE   FALSE

6. - 10.

   The date is Dec. 21. Using the earth diagram below:

   a. **draw & label** parallels **(use ruler)**;

   b. **draw** the circle of illumination;

   c. what is the sunrise direction?   E   NE   SE

   d. what is the sunset direction?   W   NW   SW

   e. Northern Hemisphere season?   _____

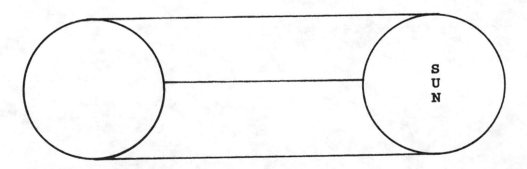

EXERCISE 2-E
EARTH-SUN RELATIONSHIPS

Student Name: _____   Lab. Sec. # _____

1. Vertical rays from the sun produce more
   heating than oblique rays;   tangent rays
   are least important.   TRUE   FALSE

2. The vertical rays from the sun are closer
   to the equator on April 1 than on August 1.   TRUE   FALSE

3. The earth is farther from the sun in
   December than in June.   TRUE   FALSE

4. The amount of daylight in the N. Hemisphere
   increases from May 1 through July 1.   TRUE   FALSE

5. In October, residents of Vincennes see sunrise in the NE, while in April they see
   sunrise in the SW.   TRUE   FALSE

6. - 10.

   The date is June 21. Using the earth diagram below:

   a. **draw & label** parallels **(use ruler)**;

   b. **draw** the circle of illumination;

   c. what is the sunrise direction?   E   NE   SE

   d. what is the sunset direction?   W   NW   SW

   e. Northern Hemisphere season?   _____

Student Name _____

EXERCISE 2-F
EARTH-SUN RELATIONSHIPS

Student Name: _____   Lab. Sec. # _____

1. There is a longer period of daylight in
    Vincennes on Dec. 25 than on Dec. 20
    every year.   TRUE   FALSE

2. The analemma is used to determine the
    angle at which the suns's rays strike
    the earth.   TRUE   FALSE

3. It is impossible for one hemisphere to
    have locations which receive 0 hrs.
    & 24 hrs. of sunlight on the same date.   TRUE   FALSE

4. From one location on earth it is possible
    to see sunrise SE and sunset NW on the
    same date.   TRUE   FALSE

5. Vincennes receives oblique & tangent rays
    at times through the day/year but never
    vertical rays!   TRUE   FALSE

6. – 10.

The date is Sept. 21. Using the earth diagram below:

   a. **draw & label** parallels **(use ruler)**;

   b. **draw** the circle of illumination;

   c. what is the sunrise direction?   E   NE   SE

   d. what is the sunset direction?   W   NW   SW

   e. Northern Hemisphere season?   _____

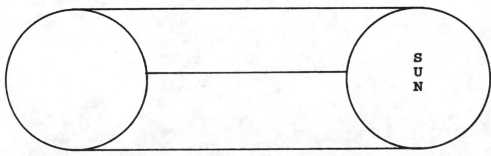

EARTH SCIENCE   LAB. EXERCISE #3

"LONGITUDE AND TIME"

OBJECTIVES:

Be Able to:

1. DETERMINE THE TIME FOR A GIVEN LONGITUDE WHEN
    PROVIDED WITH THE TIME FOR A DIFFERENT LONGITUDE.

2. DETERMINE THE LONGITUDE OF A LOCATION THROUGH
    THE USE OF A CHRONOMETER AND SEXTANT.

3. DETERMINE THE "SUN TIME" FOR A LOCATION WHEN
    GIVEN THE EXACT TIME OF ONE LOCATION, AND THE
    LONGITUDE FOR BOTH LOCATIONS.

4. EXPLAIN THE FUNCTION AND USE OF THE INTERNATIONAL
    DATE LINE.

5. EXPLAIN WHY CENTRAL TIME MERIDIANS ARE NOT
    STRICTLY UTILIZED BY COUNTRIES AND STATES.

6. DEFINE THE GLOSSARY TERMS.

NOT A BAD IDEA! FIRST WE CELEBRATE NEW YEARS OVER HERE, THEN AFTER IT'S OVER, WE COME BACK ACROSS THE RIVER WHERE THEY'RE ONE HOUR BEHIND AND CELEBRATE NEW YEARS AGAIN!

Former V.U. Student

GLOSSARY TERMS:

Military Time:

Central Time Meridian:

Sun Time:

Prime Meridian:

Greenwich Time:

Chronometer:

Zenith:

Local Time:

Sextant:

True Noon:

Day Light Saving Time:

U.S. Time Zones:

A.M. :

P.M. :

EXERCISE #3
LONGITUDE AND TIME

Time is of great influence in our daily existence. We have based our daily and yearly life styles around the concept of time and its subdivisions, but not without problems! With the speed of travel on international jet flights being so great, people who fly long distances experience jet lag and always have to reset both their watches and personal schedule to the new local time. Because the earth is round and rotates on its axis, locations around the earth are experiencing different parts of the 24 hour earth day (the approximate time required to complete one rotation on the axis of the earth) at any given instant.

The majority of people on earth rise and work with the sun and rest during the night-time hours. Thus at any given time, people at various locations around earth eat breakfast, lunch, and supper at the same instant--each in accordance with their local time. The complex aspect of this schedule is that there is no world or international time. Each country has its own time zones and most developed nations have different hours of the day but the same minutes after the hour. A universal time system will be needed in future years when space travel is more common.

An international attempt to standardize world time has resulted in the creation of 24 Central Time Meridians. These were established because the earth completes one full rotation on its axis (360°) every 24 hours. Thus 360° divided by 24 hours = 15°/hour. To double check this, if the earth turns on its axis at the rate of 15°/hour, then 24 hours x 15° = 360°.

By this system any location within 7 1/2° of longitude east or west of the closest Central Time Meridian is said to have the same time as that of the Central Time Meridian. <u>Figure 10</u> illustrates all of the Central Time Meridians around the earth. Notice that the International Date Line (180°) is shown twice, but each represents the same place.

<u>Figure 10</u> illustrates the correct use of Central Time Meridians.

*Linear Velocities of Rotation
at Various Latitudes*

Latitude degrees	Velocity miles per hour
0	1038
30	899
60	519
90	0

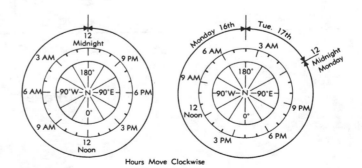

Hours Move Clockwise

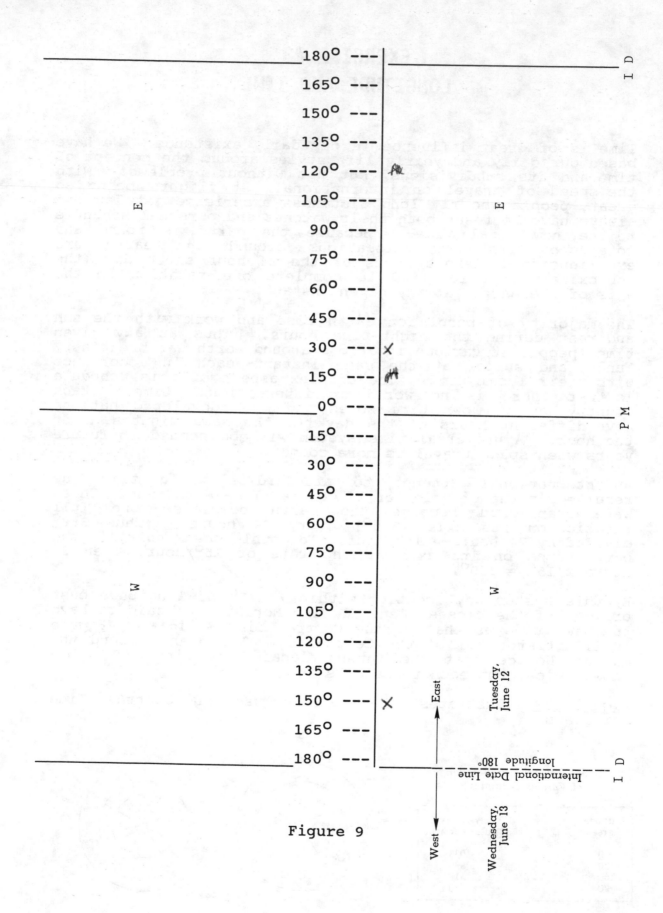

Figure 9

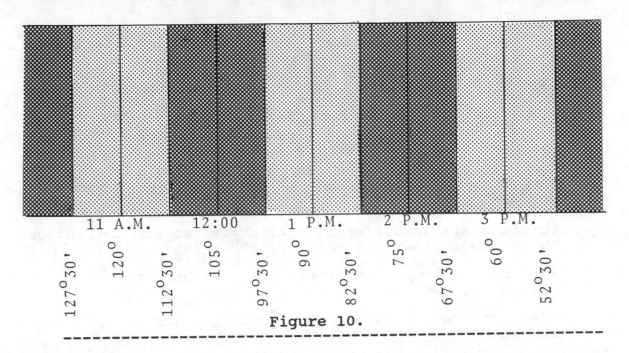

11 A.M.   12:00   1 P.M.   2 P.M.   3 P.M.

127°30' 120° 112°30' 105° 97°30' 90° 82°30' 75° 67°30' 60° 52°30'

Figure 10.

TIME PROBLEMS

Remember that the earth is rotating from west to east on its axis thus making the sun appear to rise in the east and move across the sky to set in the west. The farther one travels east therefore the later in the day the time will be. The opposite is true of course if one travels west--the farther west one travels, the earlier in the day the time will be. Learn the rules: WHEN MOVING WEST ACROSS THE EARTH'S SURFACE SUBTRACT HOURS OF TIME. WHEN MOVING EAST ACROSS THE EARTH'S SURFACE ADD HOURS OF TIME.

Time Problem Steps: (Use of Figure 9 will be helpful)

1. Convert the longitude of the location to the closest Central Time Meridian.

2. Determine the difference in degrees between locations:

   a. subtract if locations are in same hemisphere
   b. add if locations are in different hemispheres

3. Since the earth rotates 15° of longitude each hour, divide the degrees of longitude difference (step 2) by 15. This indicates the number of hours between the two locations.

4. Starting at the location of known time, take the shortest route to the location for which time is to be determined:

   a. Moving east = add 1 hr. for every 15° crossed
   b. Moving west = subtract 1 hr. for every 15° crossed.

5. When crossing the midnight line going east add 1 day; when crossing the midnight line going west subtract 1 day.

EXAMPLE 1

If it is 8:00 A.M. Tuesday at 12° E., what is the day and time at 148° W?

Steps:

1. Convert to the nearest Central Time Meridian = 15° E and 150° W.

2. Locations are in different hemispheres so add to find out how far apart they are = 15° + 150° = 165° apart.

3. Each 15° = 1 hour; therefore, $\frac{165}{15}$ = 11 hrs. time difference.

4. Direction of movement is west so hours (11) are subtracted.

   ANSWER: It is 9:00 P.M. Monday at 150° (148°) W.

EXAMPLE 2   (Crossing the International Date Line)

The International Date Line always separates two different days. This is the division where all days officially begin and end. The day always begins on the West side of the Date Line and ends on the East Side.

In crossing the International Date Line while traveling west, a day is always added but hours of time continue to be subtracted. When crossing the International Date Line while traveling east, a day is always subtraced but hours of time continue to be added. Try working longitude and time problems by crossing the Date Line and not crossing it each time to double check your answer and find the shortest route.

If it is 11:00 A.M. Wednesday at 141° E., what is the time and day at 129° W. ?

STEPS :

1. Convert to the nearest Central Time Meridian = 135° E. and 135° W.

2. Locations are in different hemispheres so add to find out how far apart they are = 135° + 135° = 270° apart.

3. Each 15° = 1 hour; therefore, $\frac{270}{15}$ = 18 hrs. time difference.

4. Direction of movement is west so hours (18) are subtracted. But instead of carrying this problem out as in example 1, cross the Date Line and the two locations are only 6 hours apart! Both locations are only 45° away from the Date Line thus 45 + 45 = 90° of longitude apart. So 90° of longitude divided by 15°/hr. = 6 hrs. of time difference.

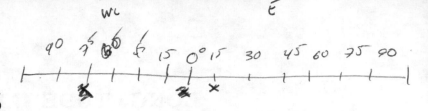

W   E
90 75 60 45 15 0° 15 30 45 60 75 90
x   x   x

PRACTICE EXERCISES

1. When it is 9:00 P.M. Monday on the Prime Meridian (0°),
    what is the time and day in Washington, D.C. (77° W.)?

    4 pm monday

2. When it is 10:00 A.M. Thursday in Berlin, Germany
    (13° E.), what is the time and day in Hong Kong, China
    (114° E.)?

    ① 120°E   ② 15)105 = 7   ④ 5 pm Thur
    - 15°E
    ─────   ③ ? 75 East = + 7 hours
      105

3. When it is 3:30 A.M. Sunday in Cairo, Egypt (32° E.),
    what is the time & day in Honolulu, Hawaii (157° W.)?

    ① 150°W   ② 15)180 = 12   ③ ? is what = -12 hrs.
    + 30°E
    ─────   3:30 SAt
      180°

4. When it is 2:25 P.M. Tuesday in Sydney, Australia
    (150° E.), what is the time and day in Bombay, India,
    (73° E.)?

       150°E   15)75 = 5   9:25 AM
      - 75°E   TUE.
      ─────
        75°

5. When it is 11:30 P.M. Wednesday in Havana, Cuba (82° W.),
    what is the time & day in Phoenix, Arizona (112° W.)?

     105°W   15)30 = 2   9:30 pm
     - 75°W   weD.
     ─────
       30°

6. When it is 7:45 A.M. Friday in Moscow, Russia (37° E.),
    what is the time & day in Tehran, Iran (52° E.)?

      45   15)15 = 1 hr   8:45 AM
     -30   FRi
     ───
      15

7. When it is 6:00 A.M. Saturday at 165° W. in the Pacific
    Ocean, what is the time & day at 75° E. in the Indian
    Ocean?

                                     10 pm
                                     SAt

61

LONGITUDE AND TIME

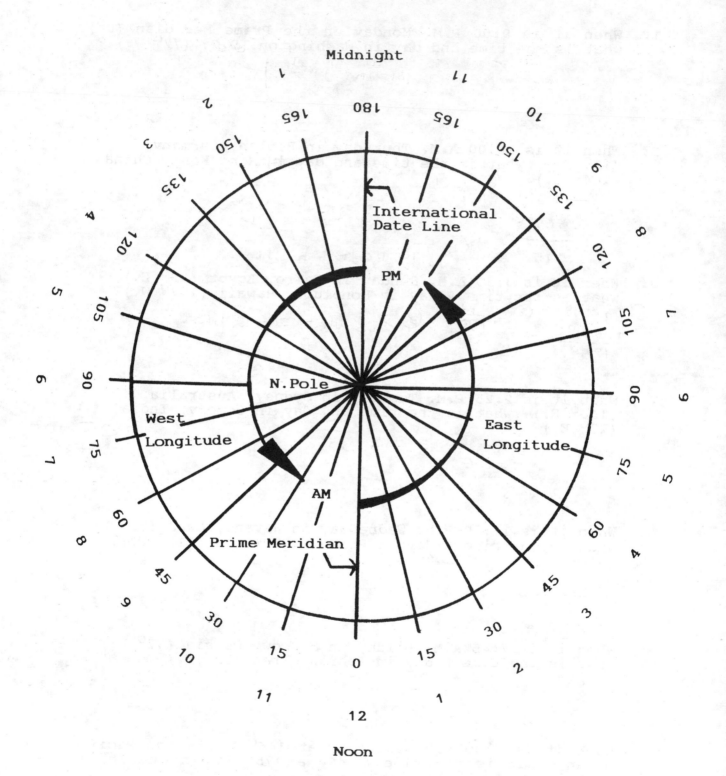

Figure 11

SUN TIME PROBLEMS

The exact local time for any given location (not rounded off to the nearest Central Time Meridian) is referred to a sun time. Sun time is determined by knowing the exact time for any longitude and the location of another place for which sun time is to be determined. Remember that the earth rotates 360° in 24 hours or 15°/each hour, or 1° every 4 minutes. (In 1 minute of time the earth would rotate through 15' of longitude). Be careful not to confuse minutes of time with ' and " of longitude! Noon local sun time refers to the period of time when the sun reaches its zenith position for that location (in other words, when the sun reaches its highest point above the horizon for that location).

STEPS:

1. Determine the difference in degrees between locations.
   NOTE: do not convert to the nearest Central Time Meridian.

   a. subtract if locations are in same hemisphere.
   b. add if locations are in different hemispheres.

2. Divide the difference in degrees by 15 and multiply any remainder by 4 (this tells you the hours & minutes the two locations are apart).

3. Starting at the location of the known sun time:

   a. if the movement is east, add the hours & minutes from step 2.
   b. if the movement is west, subtract the hours & minutes from step 2.

EXAMPLE:

If it is 5:00 P.M. Tuesday sun time at 42° W., what is the sun time at 80° W. ?

1. Both locations are in the same hemisphere so subtract:
   80° W. - 42° W. = 38° difference in location.

2. Divide 38° by 15°/hour =

$$15 \overline{)38} \quad \begin{array}{c} 2 \text{ hours} \end{array}$$
$$\underline{30}$$

   Multiply remainder by 4   8° x 4'/degree = 32 min.

   The two locations are 38° or 2 hrs. & 32 min. of time apart.

3. Direction is west so subtract hours and minutes.

   5:00 P.M.   4:60

   - 2:32   =   - 2:32
   _____   _____
                      2.28 P.M. Tuesday at 80° W.

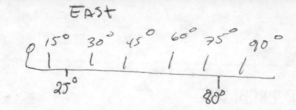

EAST

PRACTICE EXERCISES

1. If it is 10:40 P.M. Monday, <u>sun time</u> at 25° E., what is the "<u>sun time</u>" and day at Madras, India (80° E.) ?

    10:40
    3:40   + 3:40
            _____
            13:80 = 1:80 = 2:20
                    2:20 AM TUE.

2. If it is 11:25 A.M. Monday, "<u>sun time</u>" at 12° W. what is the "<u>sun time</u>" and day at Seattle, Washington (122° W) ?

    122° W
    - 12° W
    _____
    110°

    7 → 7:20
    15 ⟌110
        105
        _____
        5×4=20

    11:25
    - 7:20
    _____
    4:05 AM, MON

3. If it is 2:22 P.M. Wednesday at the Prime Meridian (0°) , what is the day and <u>sun time</u> on the International Date Line (180°) ?

    2:22 AM Thur

CHRONOMETER PROBLEMS

Another time determination that can be used is that used by navigators to determine longitude. Given the time of 2 locations and the exact longitude of one of the locations, you can determine the exact longitude of the other. To do this involves the use of an instrument known as the <u>chronometer</u> (a clock set to always read the time for the Prime Meridian -- 0 degrees longitude).

By determining when the sun reaches its zenith position at your location (use of sextant), you can determine the difference in time between your location and the Prime Meridian. Remember that the earth rotates from west to east so the sun moves westward across the sky: any location to your east is later in time and any location to your west is earlier in time.

When working these problems, if the chronometer reads P.M. your location must be in the western hemisphere, and if the chronometer reads A.M. your location must be in the eastern hemisphere. This must be true since you read the chronometer at 12:00 noon and the noon sun has either already been over the Prime Meridian or has not yet reached that point.

<u>(SEE EXAMPLES ON NEXT PAGE!)</u>

REVIEW THESE STEPS THEN COMPLETE THE PRACTICE EXERCISES

1. Chronometer is always set to give the time on the Prime Meridian.

2. Chronometer is always read when the observer sees the noon sun.

3. If the chronometer reads P.M., the observer is located in the western hemisphere; if the chronometer reads A.M., the observer is located in the eastern hemisphere.

4. If the chronometer reads A.M., subtract the reading in hours & minutes from 12:00 (11:60) your noon reading; if the chronometer reads P.M. use the reading as it is.

5. Change the hours and minutes of time into degrees & minutes of longitude:

   a. Multiply the hours by 15 (earth rotates $15°$/hour).
   b. Divide the minutes by 4 (earth rotates $1°$/4 minutes).
   c. For any remaining minutes of time:
      1 min. of time = 15' of longitude
      2 min. of time = 30' of longitude
      3 min. of time = 45' of longitude

EXAMPLE 1: If a ship's chronometer reads 6:20 A.M. as the noon sun is observed, what is the longitude of ship?

   1. Chronometer reads A.M. so location is east longitude.
   2. Subtract 6:20 from 11:60 = 11:60
   $$\underline{6:20}$$
   $$5.40 \text{ time difference}$$

   3. Multiply hrs. by 15 = 5 x 15 = $75°$
      Divide minutes by 4 = $\dfrac{40}{4}$ = $10°$

   ANSWER: $\overline{85°}$ East Longitude

EXAMPLE 2: If a ship's chronometer reads 6:25 P.M. as the noon sun is observed, what is the longitude of the ship?

   1. Chronometer indicates a P.M. reading, thus west longitude.

   2. Use exact reading of chronometer for difference = 6:25

   3. Multiply hours by 15 = 15 x 6 = $90°$
      Divide minutes by 4 = $\dfrac{25}{4}$ = $6°$

   Remainder of 1 minute   = $\underline{\hspace{1cm} 15'}$
   ANSWER: 96°   15' West Longitude

PRACTICE EXERCISES

1. A ship chronometer reads 7:30 A.M. as the noon sun is observed. What is the longitude of the ship?

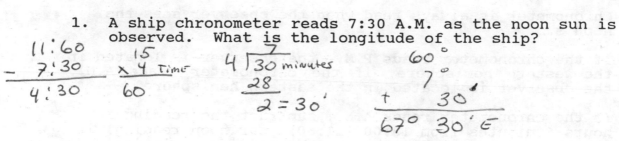

$$\begin{array}{r} 11:60 \\ -\ 7:30 \\ \hline 4:30 \end{array}$$

$$\begin{array}{r} 15 \\ \times\ 4\ \text{Time} \\ \hline 60 \end{array}$$

$$\begin{array}{r} 7 \\ 4\overline{)30}\ \text{minutes} \\ 28 \\ \hline 2 = 30' \end{array}$$

$$\begin{array}{r} 60° \\ 7° \\ +\ 30' \\ \hline 67°\ 30'\ E \end{array}$$

2. A ship chronometer reads 9:45 P.M as the noon sun is observed. What is the longitude of the ship?

9:45 pm

$$\begin{array}{r} 15 \\ \times\ 9\ \text{Time} \\ \hline 135° \end{array}$$

$$\begin{array}{r} 11° \\ 4\overline{)45}\ \text{minutes} \\ -44 \\ \hline 1 = 15' \end{array}$$

$$\begin{array}{r} 135° \\ 11° \\ 15' \\ \hline 146°\ 15'\ W \end{array}$$

3. A ship chronometer reads 4:20 A.M. as the noon sun is observed. What is the longitude of the ship?

$$\begin{array}{r} 11:60 \\ -\ 4:20 \\ \hline 7:40 \end{array}$$

$$\begin{array}{r} 15 \\ \times\ 7 \\ \hline 105 \end{array}$$

$$\begin{array}{r} 10 \\ 4\overline{)40} \end{array}$$

$$\begin{array}{r} 105° \\ +\ 10 \\ \hline 115\ E \end{array}$$

4. A ship chronometer reads 11:31 P.M. as the noon sun is observed. What is the longitude of the ship?

173° 45' W

5. Is it possible for a chronometer to read A.M. when it is P.M. in Indiana?

 yes

66

ADDITIONAL PRACTICE EXERCISES (WITH ANSWERS)
OUTSIDE CLASS ACTIVITY

1. If it is 7:00 P.M. Thursday at 60° W., what is the time and day at 90° E?

2. If it is 6:00 A.M. Saturday at 165° W., what is the time and day at 120° E.?

3. If it is 11:00 A.M. Sunday at 30° W, what is the time and day at 105° W.?

4. If it is 4:36 P.M. Tuesday at 45° W., what is the "sun time" and day at 65° E. ?

5. If it is 10:15 A.M. Friday at 150° E., what is the "sun time" and day at 145° W. ?

6. If it is 4:00 A.M. Sunday at 16° E., what is the "sun time" and day at 60° W. ?

7. A ship chronometer reads 5:30 A.M. a the noon sun is observed. What is the longitude of the ship?

8. A ship chronometer reads 8:35 P.M. as the noon sun is observed. What is the longitude of the ship?

9. A ship chronometer reads 11:59 A.M. as the noon sun is observed. What is the longitude of the ship?

ANSWERS:

1. 5:00 A.M. Friday

2. 1:00 A.M. Sunday

3. 6:00 A.M. Sunday

4. 11:56 P.M. Tuesday

5. 2:35 P.M. Thursday

6. 10:56 P.M. Saturday

7. 97° 30' E.

8. 128° 45' W.

9. 00° 15' E.

EXERCISE 3-A
LONGITUDE AND TIME

Student Name: _____ Lab. Sec. # _____

1. The earth rotates through over 100° of
    longitude in less than 10 hours.   TRUE   FALSE

2. When solving a "sun time" problem, the use
    of central time meridians is not required. TRUE   FALSE

3. 52° E. and 53° E. require the use of
    the same central time meridian.   TRUE   FALSE

4. London, England will never be earlier in
    the same day than Vincennes, Indiana.   TRUE   FALSE

5. Each new day moves from the Eastern
    Hemisphere to the Western Hemisphere.   TRUE   FALSE

6. When moving in an easterly direction
    as the International Date Line is crossed,
    1 hour is added & 1 day is subtracted.   TRUE   FALSE

7. If it is 3:30 A.M. Wednesday at 65° W.,
    what is the time and day at 17° E. ? _____

8. If it is 10:12 P.M. Monday at 85° W.,
    what is the **<u>sun time</u>** and day at 91° E.? _____

9. A ship chronometer reads 10:08 P.M. as you
    observe the noon sun. What is your
    longitude?   _____

10. When it is 4:00 P.M. Friday at 165° W.,
    what is the time and day at 159° W.? _____

EXERCISE 3-B
LONGITUDE AND TIME

Student Name: _____ Lab. Sec. # _____

1. A time zone extends 15° east and west
   of every central time meridian.   TRUE   FALSE

2. "Sun Time" calculations **do not require**
   the use of central time meridians.   TRUE   FALSE

3. All locations along the same meridian will
   have the same time regardless of latitude.   TRUE   FALSE

4. It is not possible for 2 locations within
   the same hemisphere to be 15 hours apart
   in time.   TRUE   FALSE

5. Two locations separated by 135° of
   longitude would be 7 hours apart in time.   TRUE   FALSE

6. When crossing the International Date Line
   while moving east, one hour of time is
   subtracted.   TRUE   FALSE

7. When it is 3:10 A.M. Sunday at 85° W,
   what is the time and day at 11° E.?   _____

8. If it is 4:08 A.M. "sun time" at 78° E.,
   what is the "sun time" and day at
   102° W.?   _____

9. A ship chronometer reads 4:22 A.M. as
   the noon sun is observed. What is
   the longitude of the ship?   _____

10. When it is 6:00 A.M. Monday at 150° E.,
    what is the time and day at
    105° W.?   _____

EXERCISE 3-C
LONGITUDE AND TIME

Student Name: _____ Lab. Sec. # _____

1. If 2 locations are more than 180° of longitude apart, they could not both be in the A.M.

    TRUE   FALSE

2. "Jet lag" occurs when a person gains hours in a day; this basically occurs when flying west.

    TRUE   FALSE

3. Vincennes is due north of Evansville but is an hour different in time during the winter; use of central time meridians would eliminate this.

    TRUE   FALSE

4. If it is Monday in the W. Hemisphere, it could be Sunday, but not Tuesday, in the E. Hemisphere.

    TRUE   FALSE

5. Using "sun time" calculations, 13° of longitude represents how much time?

    _____

6. How long does it take the earth to rotate through 45° of longitude?

    _____

7. When it is 4:51 A.M. Friday at 60° W., what is the time and day at 105° E.?

    _____

8. If it is 2:19 P.M. Friday at 127° W., what is the "sun time" and day at 14° W.?

    _____

9. As you observe the noon sun, the ship's chronometer reads 9:04 A.M.; the longitude is:

    _____

10. When it is 5:15 P.M. at 95° E., what is the time and day at 05° W.?

    _____

73

EXERCISE 3-D
LONGITUDE AND TIME

Student Name: _____   Lab. Sec. # _____

1. Locations that have different longitudes
    can have the same "sun time".   TRUE   FALSE

2. The only meridian which is not a straight
    line is the International Dateline.   TRUE   FALSE

3. Based on use of central time meridians,
    two locations 8° apart could both be
    at noon.   TRUE   FALSE

4. Regardless of hemisphere, as long as one
    continues to move west, time is added.   TRUE   FALSE

5. Any clock can be adjusted to become a
    chronometer.   TRUE   FALSE

6. How long does it take the earth to
    rotate through 05° of longitude?   _____

7. When it is 3:10 A.M. on Thursday at
    66° E., what is the time and day
    at 83° E.?   _____

8. If it is 4:45 A.M. Sunday at 97° E.,
    what is the time and day on the Prime
    Meridian?   _____

9. If it is 7:17 A.M. Monday at 19° E.,
    what is the "sun time" and day at
    154° W.?   _____

10. As you observe the noon sun, the
    ship's chronometer reads 11:09 A.M.;
    the longitude is: _____

EXERCISE 3-E
LONGITUDE AND TIME

Student Name: _____ Lab. Sec. # _____

1. Locations on earth cannot be more than 180° of longitude apart, thus they cannot be more than 12 hrs. apart in time.   TRUE   FALSE

2. Distance North or South of the equator has no importance in calculating time.   TRUE   FALSE

3. Using Central Time Meridians, the U.S. (including Hawaii & Alaska) stretches across more than 6 time zones.   TRUE   FALSE

4. "High Noon", a common expression in the old West, was an obvious reference to "sun time".   TRUE   FALSE

5. You live in Vincennes, IN and ask a friend living in London, England to call you at 9:00 P.M. Sunday (Vincennes time). At what time (London time) will your friend need to place the call?   _____

6. In 20 minutes the earth rotates how many degrees?   _____

7. When it is 3:10 P.M. Monday at 32° E., what is the time and day at 119° E.?   _____

8. If it is 2:10 P.M. Thursday at 147° W., what is the time and day on the Prime Meridian?   _____

9. It is 2:17 A.M. Sunday at 59° W., what is the "sun time" and day at 154° E.?   _____

10. As you observe the noon sun, the ship's chronometer reads 4:15 P.M., the longitude is?   _____

EXERCISE 3-F
LONGITUDE AND TIME

Student Name: _____ Lab. Sec. # _____

1. The United States can experience 2 but
    not 3 different days at the same time.   TRUE   FALSE

2. Hawaii will always celebrate the exact
    arrival of each new year later than
    Japan.   TRUE   FALSE

3. If your shadow is cast to the West,
    you are in the A.M. portion of your
    day.   TRUE   FALSE

4. Convenience and population needs
    prevent the universal use of Central
    Time Meridians.   TRUE   FALSE

5. The sun moves across the sky at the
    rate of _____° per hour.

6. Explain the saying from the 1800's,
    "the sun never sets on the British Empire"!

    _____

    _____

7. When it is 10:10 P.M. Saturday at 7° E.,
    what is the time and day at 151° E.?   _____

8. If it is 5:15 P.M. at 27° W, what is
    the time and day on the Prime Meridian?   _____

9. It is 5:53 A.M. Friday at 71° W, what
    is the "sun time" and day at 140° W.?   _____

10. As you observe the noon sun, the ship's
    chronometer reads 7:15 P.M.; the
    longitude is: _____

EARTH SCIENCE   LAB. EXERCISE #4

"LAND SURVEY SYSTEMS"
- -

OBJECTIVES:

BE ABLE TO:

1. DISCUSS THE HISTORY OF DEVELOPMENT OF THE LAND
    SURVEY SYSTEM.

2. IDENTIFY AND EXPLAIN METES AND BOUNDS, AND THE
    FRENCH "LONG LOTS" SYSTEMS.

3. DETERMINE EXACT TOWNSHIP & RANGE, SECTION, QUARTER
    AND SMALLER DIVISIONS OF LAND ACCORDING TO THE
    TOWNSHIP & RANGE SYSTEM.

4. DETERMINE THE SIZE OF DIVISIONS WITHIN THE TOWNSHIP
    AND RANGE SYSTEM IN TERMS OF ACRES AND FRACTIONS
    OF ACRES.

5. IDENTIFY AND USE THE LAND SURVEY SYSTEM ON TOPO-
    GRAPHIC MAPS.

6. DETERMINE ACREAGE OF SUBDIVISIONS OF SECTIONS.

7. DETERMINE SECTION NUMBERS (1-36) WITHIN A CON-
    GRESSIONAL TOWNSHIP.

8. ESTIMATE WITH REASONABLE ACCURACY THE ACREAGE
    OF IRREGULAR SHAPED LAND AREAS OR LAKES.

9. EXPLAIN THE SIGNIFICANCE OF THE "STANDARD PARALLELS"
    WITHIN THE TOWNSHIP AND RANGE SYSTEM.

10. DEFINE AND/OR EXPLAIN THE GLOSSARY TERMS.

B.C. by permission of Johnny Hart and Field Enterprises, Inc.

GLOSSARY TERMS:

Metes & Bounds:

French Plotting System:

French Long Lots:

Dead Reckoning:

Township:

Range:

Principal Points:

Principal Meridians:

"Gore District" of Indiana:

Tier:

Base Lines:

Congressional Township:

Section:

Quarter-Sections:

Acre:

Standard Parallel:

Convergence Factor:

Platt Book:

EXERCISE #4
LAND SURVEY SYSTEMS

This exercise will examine the history and methods of the different types of land survey used in the United States.

LAND SURVEY SYSTEMS

Three land survey systems can be seen today in the United States:

Metes and Bounds
French Plotting System
Township Metes and Bounds.

The oldest system first used when the North American continent was colonized during the 17 century was called "Metes and Bounds". In portions of the United States such as Colonial America, Texas, and Kentucky this system is still evident. By this system an outstanding physical feature such as a large boulder or tree was selected by the surveyer as a reference point along the perimeter of one's property and then simple compass readings and distance (often pacing) were used to trace the boundaries. This system was unsatisfactory because of the common selection of movable or temporary objects as reference points, plus the human error factor in reading a simple compass and other inexact measurement techniques. Areas plotted in this manner that have not been resurveyed by a different method retain their highly irregular shapes and sizes; no two areas are exactly alike. Following roads in areas where the Metes and Bounds method was used is far different from the network of straight roads found in this region where a different system was adopted. (note this on the map of a portion of SW Indiana in Figure 12).

FRENCH PLOTTING SYSTEM

This surveying system is common to Knox and Floyd counties in Indiana, and also portions of Louisiana. On this system, land was divided into rectangular blocks or tracts using rivers and an end boundary (most often aligned in a NW-SE direction). The idea was to provide each settler (usually a farmer) with a portion of the river to ship his produce; bottom land and higher terrace land enabled farming different crops, and higher ground for a safe location of his home. The wealther the farmer, the wider his strip of land. Study this system on the maps of Vincennes, IN & IL and Thibodaux, LA. on the following pages; you may also wish to examine the large wall map of Knox County provided in the Earth Science Lab. to see the French "long lots".

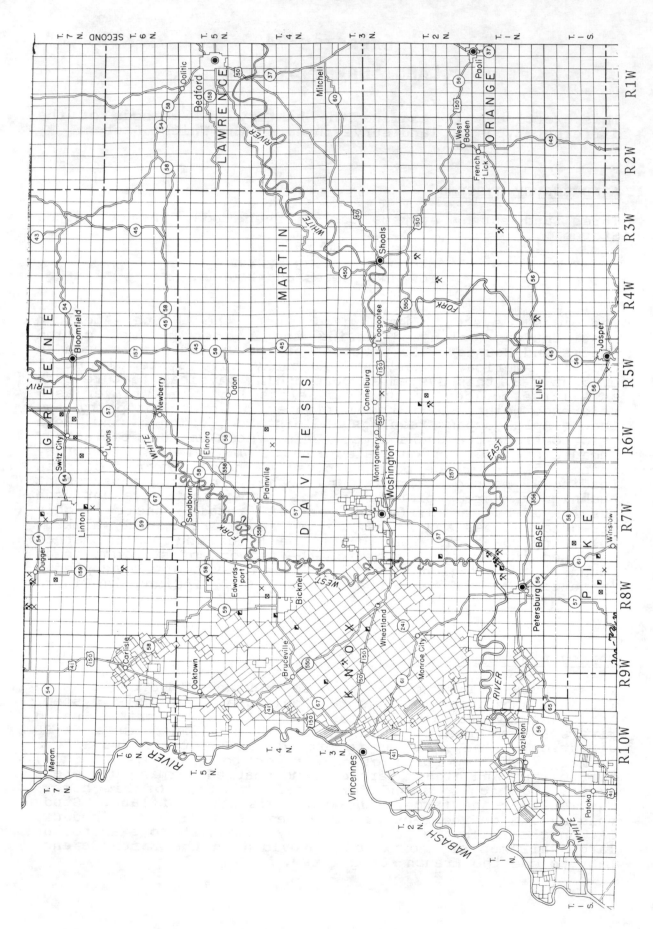

Figure 12.

84

LAND SURVEY SYSTEMS ON TOPOGRAPHIC MAPS

Take a few minutes to examine the following topographic maps provided by your lab instructor:

    A.   Oolitic, Indiana

    B.   Bright Angel, Arizona

    C.   Princeton, Indiana - Illinois

1.   Can you see any evidence of land surveying on any of the 3 maps?

2.   If so, which maps?

3.   Which map has the greatest amount of "confusion" or irregularity in division of land into measured units?

4.   What does the presence of metes & bounds and/or French long lots indicate about the history of settlement of a region?

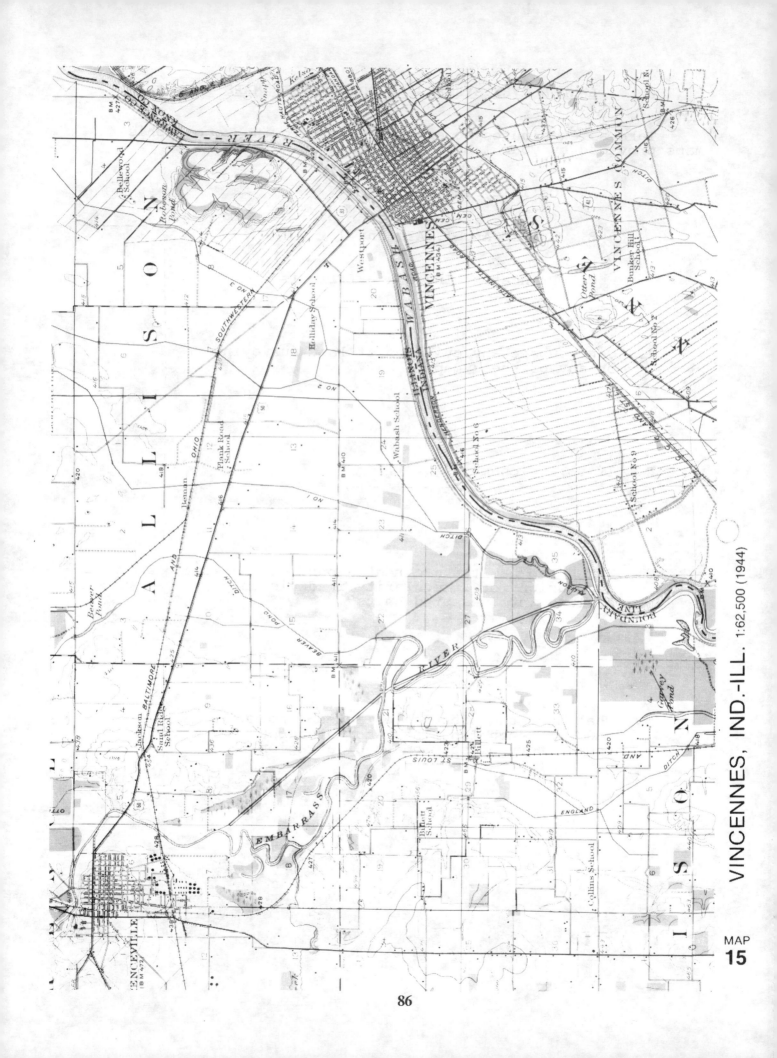

VINCENNES, IND.-ILL. 1:62,500 (1944)

MAP
15

86

Vincennes, Ind.-Ill. 15

1:62,500 (1944)

This area is shown on the Vincennes (NJ 16-5) 1:250,000 sheet. It is also covered by 7½' quadrangles published in 1961.

A French fort and fur trading post was established in 1732 at Vincennes to guard the Wabash River route between Lake Erie and the Ohio River. The French settlers were primarily interested in fur trading, but each had a strip of cultivable land on one of the prairies near the settlement, and the right to pasture livestock on the Common to the south. American authorities did their best to respect French property rights when the remainder of the area was divided according to the congressional land survey system. The wide floodplain of the Wabash River has been drained for Corn Belt farming and Vincennes, which was the first capital of the Indiana Territory, has become a county seat and service center for a prosperous agricultural area.

Noteworthy features

- Contrasting French (long lots, Common) and congressional land survey systems.
- Grid street pattern of Vincennes oriented toward the river.
- Roads and railroads attracted by the early settlement.
- Water control works (levees, drainage ditches) on the floodplains of the Wabash and Embarrass rivers.
- Plank Road School.
- Oil storage tanks south of Lawrenceville.

Climate

	J	F	M	A	M	J	J	A	S	O	N	D	Year
T.	33	36	44	56	66	75	79	77	70	59	45	35	56
P.	3.6	2.8	4.0	4.4	4.7	4.3	3.8	3.3	3.6	2.7	3.6	3.1	43.7

Population

	1970	1960	1950	1940	1930	1920	1910	1900
Vincennes	19,867	18,046	18,831	18,228	17,564	17,160	14,895	10,249
Lawrenceville	5,863	5,492	6,328	6,213	6,303	5,080	3,235	1,300

Scale

SCALE 1:62500

CONTOUR INTERVAL 20 FEET
DATUM IS MEAN SEA LEVEL

THIBODAUX, LA. 1:62,500 (1962)

MAP
9

Thibodaux, La. 9

1:62,500 (1962)

French settlers divided much of southern Louisiana into long narrow properties more or less at right angles to one of the major streams. The natural levees alongside the rivers were the highest and best drained portions of the low-lying delta country swamplands, and settlement hugged the banks of the streams. The upper parts of the levee backslopes have been cleared and drained for production of sugarcane and other crops, but vast areas of backswamp remain undrained. The discovery of oil and gas fields has given some backswamp areas new economic importance in the twentieth century.

This area is shown on on the New Orleans (NH 15-9) 1:250,000 sheet. It is also covered by 7½' quadrangles published in 1938 and revised in 1962.

Noteworthy features

- French long lot system of land survey, with rectangular survey of the unoccupied backswamps.
- Concentration of houses and highways on the natural levees.
- Sugar plantation complexes, some of which even have their own railroad sidings, scattered through the fields.
- Canals, ditches, rectangular contour lines, and other signs of extensive drainage works.
- Relationship between drainage and cleared land.
- Oil and gas fields and associated transport facilities, such as canals, pipelines, and light-duty roads.
- Mixture of French and English place-names, with such local terms as bayou and parish.

Climate

Donaldsonville (35 miles north of Thibodaux)

	J	F	M	A	M	J	J	A	S	O	N	D	Year
T.	56	58	62	69	76	81	82.	83	79	71	61	57	70
P.	4.8	5.1	5.1	5.1	5.7	4.6	6.2	5.7	5.4	2.7	4.3	5.1	59.9

Population

	1970	1960	1950	1940	1930	1920	1910	1900
Thibodaux	14,925	13,403	7,730	5,851	4,442	3,526	3,253	2,078

Scale

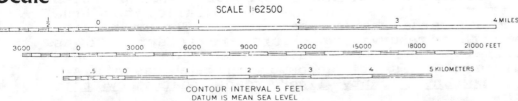

SCALE 1:62500

CONTOUR INTERVAL 5 FEET
DATUM IS MEAN SEA LEVEL

Map 9 and this page, Reprinted by permission from Karl B. Raitz & John Fraser Hart, CULTURAL GEOGRAPHY ON TOPOGRAPHIC MAPS (John Wiley & Sons Publishing Co., 1975).

CONGRESSIONAL TOWNSHIP & RANGE SYSTEM

The majority of the United States, has been surveyed by the Township and Range System which was legislated by the United States government in 1785. Exceptions to this system still exist where earlier developments had previously been surveyed (i.e. metes and bounds, or the French plotting system).

In the Township and Range System, divisions of land are established by a series of N-S and E-W lines thus creating a grid network based on the latitude and longitude system. Within this series of N-S and E-W running lines, certain lines (parallels and meridians) are selected as reference points. The reference parallels are called "base lines", and the reference meridians are called "principal meridians". Remember the latitude and longitude orientation--parallels (base lines) run E-W and measure N-S; meridians (principal meridians) run N-S and measure E-W.

The base lines and principal meridians running across the United States have been given names or numbers (see Figure 13), and within this system additional lines are established every 6 miles N-S and E-W thus creating a mosaic of squares 6 miles by 6 miles. These tracts of land (36 square miles in size) are called <u>Congressional Townships.</u> **NOTE:** Our political or civil townships do not necessarily coincide with the congressional townships of the land survey system. **<u>DO YOU HAVE ANY IDEA WHY?</u>**

The rows or tiers of townships created by the grid network are numbered North or South of a Base Line and East or West of a Principal Meridian (shown in <u>Figure 14</u> Thus by referring to township and range readings, a congressional township's location can be exactly pinpointed (townships are numbered North and South of base lines; ranges are numbered East and West of principal meridians). As in latitude and longitude readings, the N-S location (township) is always given first! **<u>EXAMPLE:</u>** the township marked X in <u>Figure 14</u> would correctly be identified as T. 3 S., R. 6 W.

Notice also how the fifth tier of townships in <u>Figure 14</u> has the range lines offset to the East and West of the range lines directly below them. This occurs because of a resurveying to offset the diminishing size of townships which would result if range lines continued to parallel the true meridians which converge nearer the poles. The correction parallels where this resurveying occurs are referred to as "standard parallels" and N-S roads often jog at these correction points (90° is not uncommon in many cases!) <u>Are you familiar with any "county line" roads that show this feature?</u>

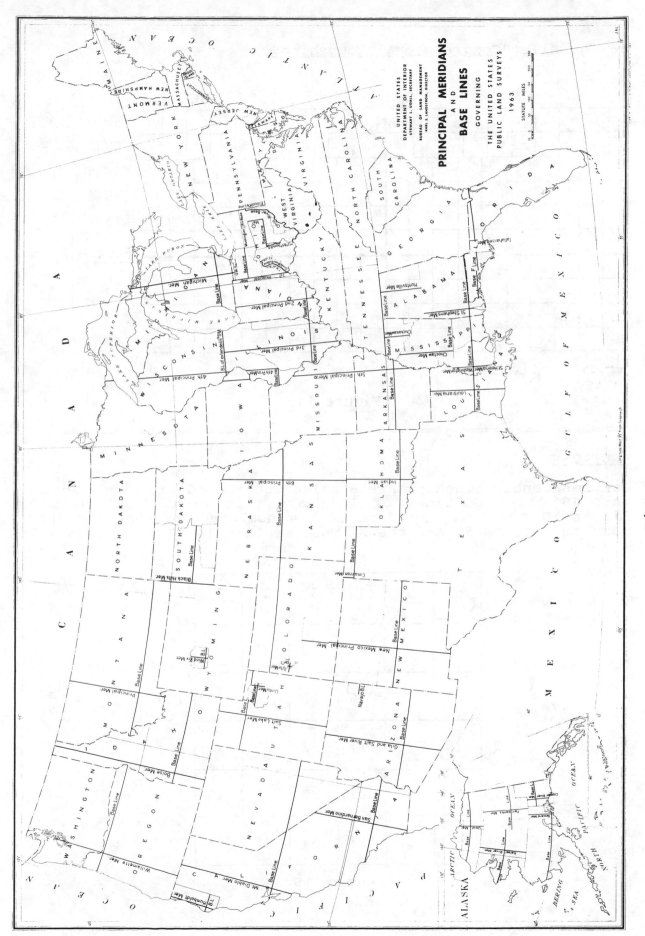

PRINCIPAL MERIDIANS
AND
BASE LINES

GOVERNING

THE UNITED STATES
PUBLIC LAND SURVEYS

1963

UNITED STATES
DEPARTMENT OF INTERIOR
STEWART L. UDALL, SECRETARY

BUREAU OF LAND MANAGEMENT
KARL S. LANDSTROM, DIRECTOR

STATUTE MILES

Figure 13

Reprinted courtesy U. S. Geological Survey.

36 Sq. Mile "Congressional Townships"

6 mi × 6 mi →

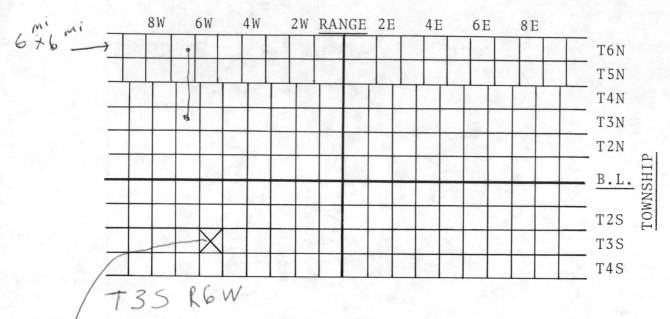

T3S R6W

Figure 14.

===

SECTIONS

Congressional Townships are 6 miles square and are subdivided into 36 equal sections of 1 square mile each. Notice in Figure 15 how townships are always numbered so that the NE most section is #1 and the SE most section is #36.

	31	32	33	34	35	36	
1	6	5	4	3	2	1	6
12	7	8	9	10	11	12	7
13	18	17	16	15	14	13	18
24	19	20	21	22	23	24	19
25	30	29	28	27	26	25	30
36	31	32	33	34	35	36	31
?							?

1 mi × 1 mi = 640 ACRES

Figure 15.

"ONE TOWNSHIP"

In reference to Figure 15, each of the square mile sections contains 640 acres. Sections can also be divided into quarter sections of 160 acres and would be referred to as either the NE1/4, NW1/4, SE1/4, or the SW1/4 of the section. Figure 16 illustrates the subdivision of a section into the 4 quarter sections.

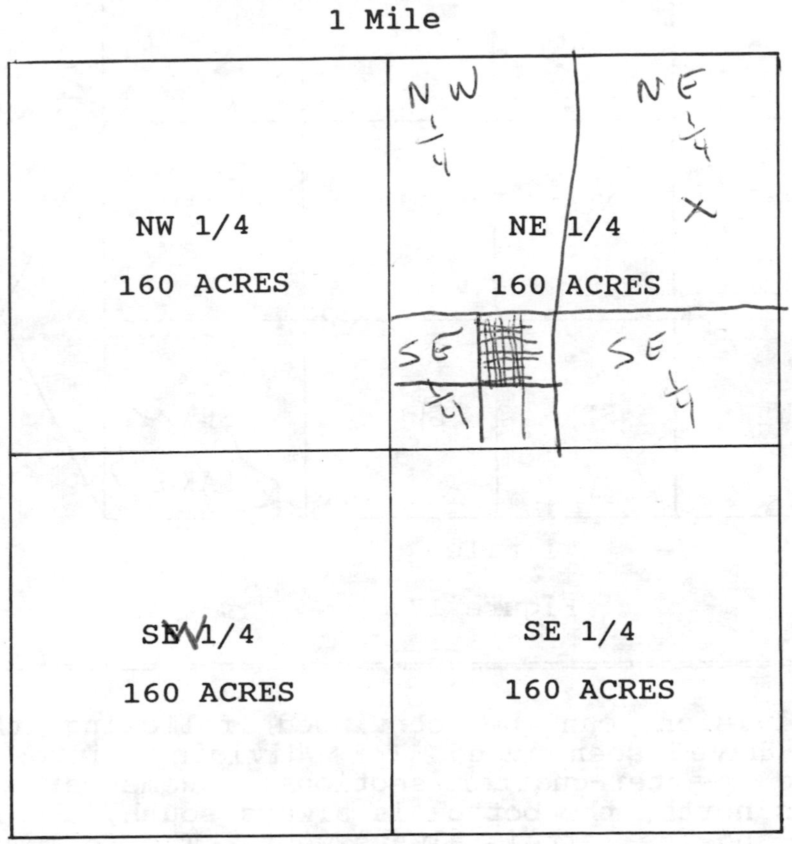

1 Mile

NW 1/4 160 ACRES	NE 1/4 160 ACRES
SW 1/4 160 ACRES	SE 1/4 160 ACRES

Figure 16.

==

The quarter sections shown above in Figure 16 can also be further subdivided into what are referred to as quarter-quarter sections which would of course contain 40 acres each (1/4 of 160). They are also named as the NE1/4, NW1/4, SE1/4, or the SW1/4 of the quarter section of which they are a part. Figure 17 on the following page illustrates the subdivision into quarter-quarter sections.

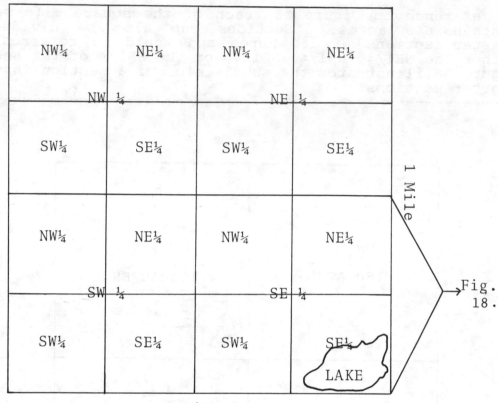

NW¼	NE¼	NW¼	NE¼
NW ¼		NE ¼	
SW¼	SE¼	SW¼	SE¼
NW¼	NE¼	NW¼	NE¼
SW ¼		SE ¼	
SW¼	SE¼	SW¼	SE¼ LAKE

1 Mile

→ Fig. 18.

1 mile

Figure 17.

==

Further subdivision can be continued following the same process you have seen used for dividing into quarter sections, and quarter-quarter sections. Remember that the top is always north, the bottom is always south, the right is always east, and the left is always west. The following page contains some examples of the correct division of a section into smaller units.

In Figure 18, we can see that the SE 1/4 of the section shown in Figure 17 has been isolated and enlarged, and then divided into quadrants. The center of the lake, for example, is located in the SE1/4 of the SE1/4 of the SE1/4 of the section. The number of the section and the location of the section in terms of Township and Range would also be given. (Example SE1/4, SE1/4, SE1/4 of Section 14, Township 7 N. Range 4 E.). The approximate size of the lake shown in Figure 17 and 18 would be 15 acres.

Study the examples of section subdivision which follow; note both the location description and acreage; then proceed to the practice exercises.

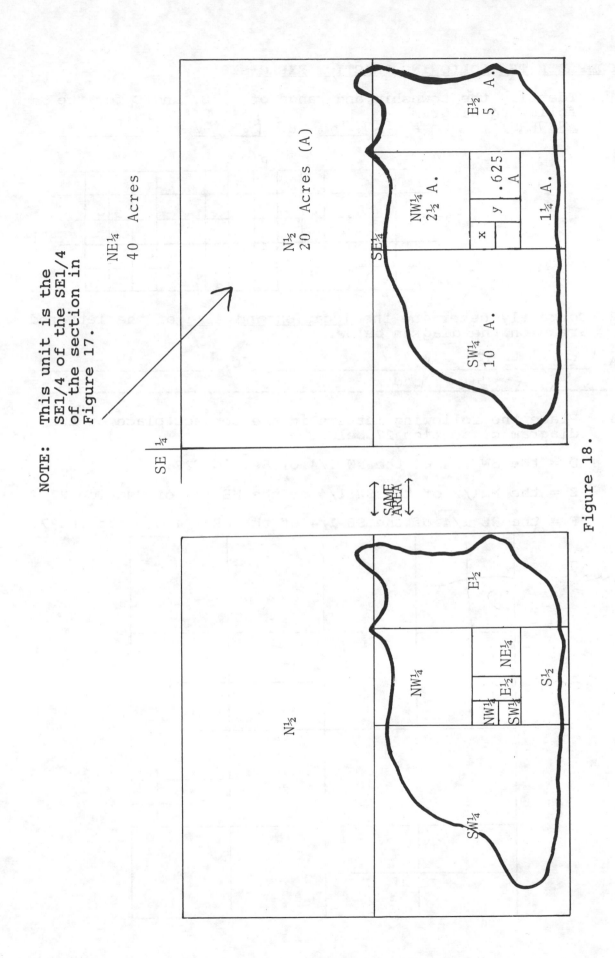

NOTE: This unit is the SE1/4 of the SE1/4 of the section in Figure 17.

NE¼   40   Acres

N½   20   Acres   (A)

NW¼   2½ A.

SE¼

x   y   .625 A.

1¼ A.

E½   5   A.

SW¼   10   A.

SE ¼

SAME AREA

N½

NW¼

SW¼

E½

NW¼   E½   NE¼
SW¼

S½

Figure 18.

95

COMPLETE THE FOLLOWING PRACTICE EXERCISES

1. Identify the township and range of A, B, and C for the following diagram.
ANSWERS: A. T2N R5W B. T1S R3W C. T2N R7E

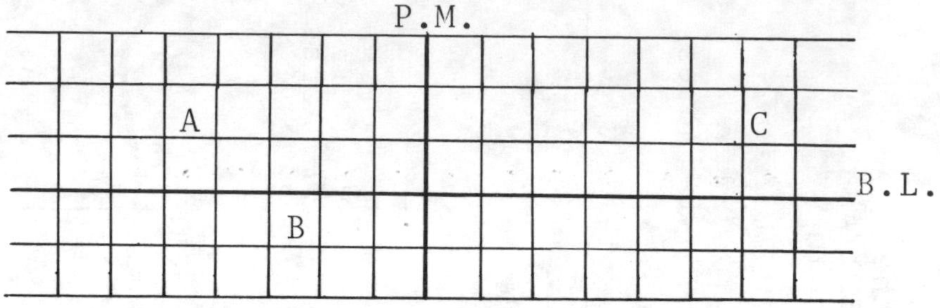

2. Correctly determine the location and size of the lettered areas on the diagram below.

A. SE 1/4, SE 1/4 ~~100~~ B.________ C.________

NW 1/4 SE 1/4 SEC 27 40 ACRES

---

3. Place the following letters in the correct place on the diagram of section 27 below.

D = the SW 1/4 of the NW 1/4 of Section 27.

E = the N 1/2 of the SW 1/4 of the NE 1/4 of Section 27.

F = the NE 1/4 of the SE 1/4 of the NE 1/4 of Section 27.

SE 1/4
B. NW 1/4, NE 1/4 sec 27
40 ACRES.

C. SW 1/4, SE 1/4
sec 27 10 ACRES.

SW 1/4 SE 1/4
NE 1/4 SW 1/4

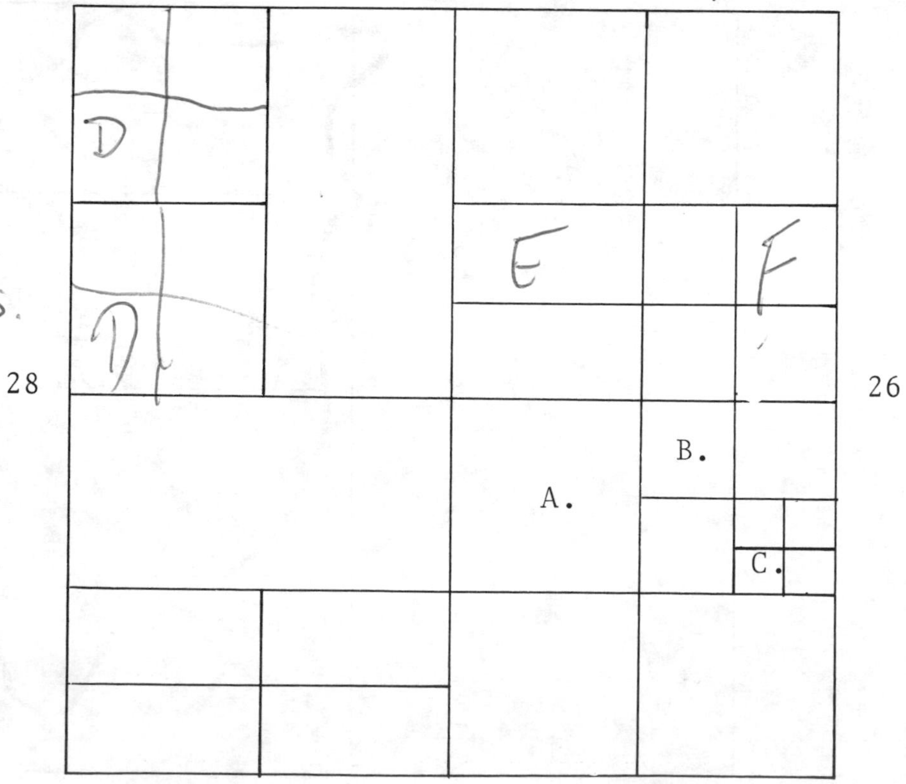

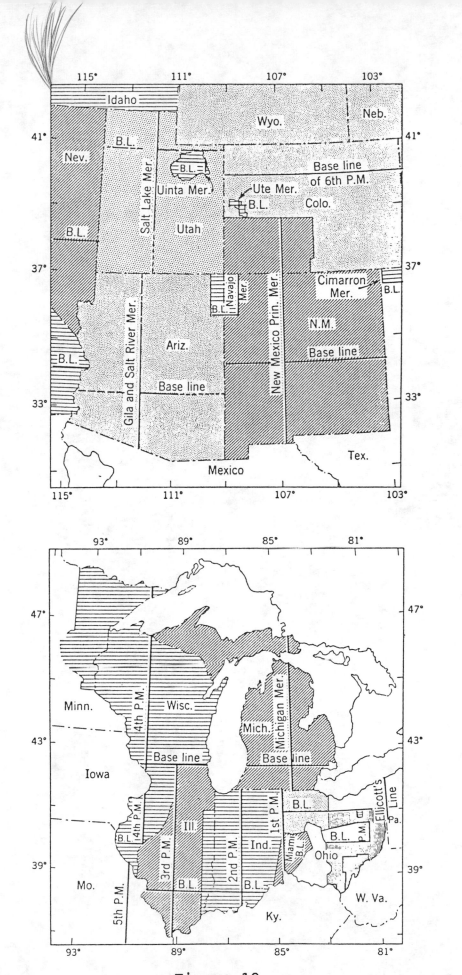

Figure 19.

EXERCISE 4-A
LAND SURVEY SYSTEMS

Student Name: _____   Lab. Sec. # ____

1. Portions of Indiana have been
    surveyed by both the Township &
    Range, and the French Plotting
    System.   TRUE   FALSE

2. A Congressional Township consists of
    one section; this section is one
    mile square or 640 acres.   TRUE   FALSE

3. Both a Principal Meridian and Base Line
    of the Land Survey System pass through
    Indiana.   TRUE   FALSE

4. A section could have a number of 25 but
    congressional townships never are numbered
    higher than 5.   TRUE   FALSE

5. Moving due east for 4 miles from the center
    of section 3 would place you in the center
    of section _____.

6. A congressional township contains ____ sq.
    miles.

7. Because of convergence of meridians, range
    lines are off-set by resurveying every ____
    miles.

**USE THE SECTION DIAGRAM ON THE REVERSE SIDE TO CORRECTLY
LOCATE EACH OF THE FOLLOWING:**

8. E 1/2 of the NW 1/4 of the Section 10.

9. SW 1/4 of the SW 1/4 of the NE 1/4 of the Section 10.

10. Draw a 20 acre lake in the NE 1/4 of the Section 10.

3

9

EXERCISE 4-B
LAND SURVEY SYSTEMS

Student Name: _____ Lab. Sec. # _____

1. Throughout much of the Eastern United
   States, the divisions of metes and
   bounds can still be seen.   TRUE   FALSE

2. One-half of a quarter section equals
   40 acres.   TRUE   FALSE

3. The Congressional Township system is
   more visible and used in W. U.S.A.
   than the East.   TRUE   FALSE

4. Some topographic maps contain the Township and Range system as well as latitude
   and longitude.   TRUE   FALSE

5. The most common method of topographic map
   construction is the field survey method.   TRUE   FALSE

6. Both a Base Line and Principal Meridian   TRUE   FALSE
   pass through Indiana.

7. Moving due south for 8 miles from the
   center of section 3 would place you in
   the center of section _____.

**USE THE SECTION DIAGRAM ON THE REVERSE SIDE TO CORRECTLY
LOCATE EACH OF THE FOLLOWING:**

8. W 1/2 of the NW 1/4 of Section 14.

9. NW 1/4 of the SW 1/4 of the SW 1/4 of Section 14.

10. Draw a 10 acre lake in the SE 1/4 of the NW 1/4 of
    Section 14.

11

15

EXERCISE 4-C
LAND SURVEY SYSTEMS

Student Name: _____ Lab. Sec. # ____

1. Metes and bounds surveying results
   in lots that are very geometric only.   TRUE   FALSE

2. Political townships vary in size;
   congressional townships do not.   TRUE   FALSE

3. Township refers to North and South while
   Range refers to East and West.   TRUE   FALSE

4. At a location of Township 4 N., you would
   be 24 miles maximum away from the base
   line.   TRUE   FALSE

5. Moving due West 10 miles from the center
   of Section 3 would place you in the center
   of Section _____.

6. It is ____ miles from the center to edge of
   a Congressional Township.

7. One-half of a quarter-quarter section would
   contain how many acres?   _____

**USE THE SECTION DIAGRAM ON THE REVERSE SIDE TO CORRECTLY
LOCATE EACH OF THE FOLLOWING:**

8. NE 1/4 of the NE 1/4 of Section 8.

9. E 1/2 of the SW 1/4 of the SE 1/4 of Section 8.

10. Draw a 10 acre lake in the SW 1/4 of the SE 1/4
    of the NW 1/4 of Section 8.

EXERCISE 4-D
LAND SURVEY SYSTEMS

Student Name: _____ Lab. Sec. # _____

1. French "long lots" are most common
   in areas on or near rivers.   TRUE   FALSE

2. To walk around the perimeter of an
   entire congressional township is a
   trip of <u>over</u> 36 miles.   TRUE   FALSE

3. One-half of a Congressional Township
   consists of 18 sections or 11,520 acres.   TRUE   FALSE

4. Township 3 North begins _____ miles
   from its base line.

5. Moving due West for 16 miles from the
   center of Section 3 would place you in
   the center of Section _____.

6. It is ____ miles around the perimeter of
   a section.

7. A quarter, quarter, quarter section contains
   _____ acres.

<u>USE</u> <u>THE</u> <u>SECTION</u> <u>DIAGRAM</u> <u>ON</u> <u>THE</u> <u>REVERSE</u> <u>SIDE</u> <u>TO</u> <u>CORRECTLY</u>
<u>LOCATE</u> <u>EACH</u> <u>OF</u> <u>THE</u> <u>FOLLOWING:</u>

8. S 1/2 of the NE 1/4 of Section 9.

9. NW 1/4 of the NE 1/4 of the SW 1/4 of Section 9.

10. Draw a 5 acre lake in the NW 1/4 of the SE 1/4 of
    the SW 1/4 of Section 9.

4

8

EXERCISE 4-E
LAND SURVEY SYSTEMS

Student Name: _____ Lab. Sec. # _____

1. The Congressional Land Survey System
    of Township and Range is over 200 years
    old.   TRUE   FALSE

2. It is farther from the top to bottom of a
    Congressional Township than around the
    perimeter of a section.   TRUE   FALSE

3. A Congressional Township is square and
    contains _____ sections which are 1 sq.
    mile each.

4. Range 5 W begins _____ miles from its
    Principal Meridian.

5. Moving due east for 10 miles from section
    21 would place you in the center of Section
    _____.

6. A Congressional Township contains ___ sq.
    miles while a quarter section contains
    ____ acres.

7. A farmer says he "has to plow the back 40";
    what fraction of a section of land is that
    unit?   _____

**USE THE SECTION DIAGRAM ON THE REVERSE SIDE TO CORRECTLY
LOCATE EACH OF THE FOLLOWING:**

8. SE 1/4 of the SE 1/4 of the SE 1/4 of the Section 21.

9. N 1/2 of the SW 1/4 of the Section 21.

10. Draw an 80 acre lake in the NW 1/4 of the Section 21.

16

20

EXERCISE 4-F
LAND SURVEY SYSTEMS

Student Name: _____ Lab. Sec. # ____

1. Metes and Bounds surveying was more
    common in Western USA than any other
    region.   TRUE   FALSE

2. It is possible for a section of land
    (i.e. section 36) to be part of 2
    separate Congressional Townships.   TRUE   FALSE

3. If you are located within township 4 S,
    you could be over 24 miles from your
    Base Line.   TRUE   FALSE

4. It is possible to see 36 sections but not
    a complete Congressional Township on a
    topographic map.   TRUE   FALSE

5. Moving due north for 12 miles from the
    center of Section 32 would place you in
    the center of Section _____.

6. A section contains ____ acres or ____ sq.
    miles.

7. The majority of Lawrenceville, Illinois is
    located within Section _____. (See p. 86).

**USE THE SECTION DIAGRAM ON THE REVERSE SIDE TO CORRECTLY
LOCATE EACH OF THE FOLLOWING:**

8. SE 1/4 of the NE 1/4 of the SW 1/4 of the Section 34.

9. W 1/2 of the SW 1/4 of the Section 34.

10. Draw a 60 acre lake in the W 1/2 of the SW 1/4 of
    the Section 34.

27

33

110

EARTH SCIENCE   LAB. EXERCISE #5
"MAP READING"

OBJECTIVES:

Be Able to:

1. EXPLAIN THE DIFFERENCE BETWEEN TOPOGRAPHIC AND
   PLANIMETRIC MAPS

2. DESCRIBE THE CONSTRUCTION OF TOPOGRAPHIC MAPS BY
   FIELD SURVEYING METHODS.

3. DISCUSS THE CONSTRUCTION OF TOPOGRAPHIC MAPS
   THROUGH THE USE OF AERIAL PHOTOGRAPHS.

4. EXPLAIN AND DETERMINE THE DIFFERENCE IN LARGE
   AND SMALL SCALE MAPS.

5. DESCRIBE AND USE VERBAL, GRAPHIC, AND REPRESENTATIVE
   FRACTION MAP SCALES.

6. IDENTIFY TOPOGRAPHIC MAP SYMBOLS USING THE U.S.G.S.
   SYMBOLS SHEET IN THE LAB MANUAL.

7. ASSOCIATED THE FIVE (5) STANDARD COLORS USED ON
   TOPOGRAPHIC MAPS WITH THEIR CORRECT FEATURES.

8. DETERMINE THE AREA IN SQUARE MILES COVERED BY
   TOPOGRAPHIC MAPS.

9. EXPLAIN AND DETERMINE THE ANGLE OF MAGNETIC
   DECLINATION FOR ANY TOPOGRAPHIC MAP.

10. USE LATITUDE AND LONGITUDE COORDINATES TO
    DETERMINE EXACT LOCATIONS ON TOPOGRAPHIC MAPS.

11. GIVE DETAILS ON HOW AND WHERE TO PURCHASE TOPO-
    GRAPHIC MAPS; ALSO HOW TO USE A STATE INDEX
    MOSAIC IN ORDERING.

12. DEFINE AND/OR EXPLAIN THE GLOSSARY TERMS.

GLOSSARY TERMS:

Index Mosaic:

15' Series:

7 1/2" Series:

Map Area:

U.S.G.S.:

Quadrangle:

1: 1,000,000

1: 62,500

1: 48,000

1: 24,000

Field Surveying:

Stereo Pairs:

Stereoscope:

Planimetric:

Topographic:

Relief:

Photo Revision:

Aerial Photography:

Map Scale:

Declination:

Large Scale Map:

Small Scale Map:

Datum Plane:

Elevation:

Field Check:

Bench Mark:

EXERCISE #5

FUNDAMENTALS OF MAP READING

This laboratory exercise will introduce a number of basic skills for use in the interpretation of data found on "topographic" maps. Included in this exercise will be discussion of the following:

- A. Topographic vs. Planimetric Maps
- B. Construction of Topographic Maps
- C. Map Scale
- D. Determination of Map Area in Sq. Miles
- E. Angle of Declination
- F. U.S.G.S. Topographic Map Symbols
- G. Use of Colors on Topographic Maps
- H. Application of Latitude-Longitude Coordinates
- I. Topographic Map Selection/Purchase Information

Most of the maps we frequently use are "planimetric"; that is, they are two dimensional. While showing distances N-S and E-W, no vertical measurement (relief) is given. State highway maps and the maps of an atlas are typical planimetric maps. If a map shows elevations, thus a 3rd. dimension, it is referred to as a "topographic map". Topography is the "relief" shown by the earth's surface; the study of the earth's topography is a branch of Geology known as "Geomorphology". Since topographic maps show the elevations of land, they are of great importance and use to many people (geologists, geographers, military troops, outdoor enthusiasts, and landowners).

In addition to the elevations shown on topographic maps (commonly referred to as topo's.), additional attention is also given to physical features such as mountains, rivers, lakes, depressions, degree of slope and so forth. With skill it is possible to determine such information as direction and rate of stream/river flow, lake depths, which routes present the easier or more difficult hike, and recognition of landmarks.

Elevation, thus relief, is shown on topographic maps by means
of "contour lines". The following terms are important to
fully understand the concept of topographic maps:

DATUM PLANE: A frame of reference from which
                elevations are measured. The datum
                plane for nearly all U.S.G.S. maps
                is mean sea level. Negative
                elevations can also be shown.

ELEVATION: The vertical distance above or be-
                low a datum plane in feet or meters.
                Selected locations have exact
                elevations plotted; brass markers,
                known as "bench marks", can be
                found at these locations with the
                following appearance and information.
                On a topographic map the location &
                elevation of a bench mark appears like
                this: BM X 4000

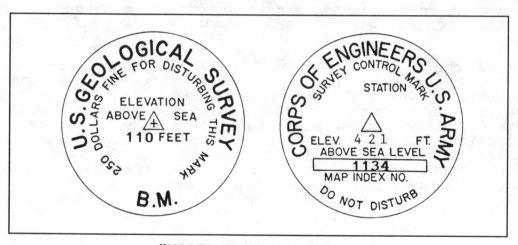

"MODEL BENCH MARK"

CONTOUR LINE: A brown line drawn across a topographic
                map which connects points of exactly the
                same elevation; consequently, each
                distinct contour line represents a "plane"
                across the earth's surface parallel
                to sea level. Contour lines will
                indicate the 3-D appearance of an area
                when properly read!

116

TOPOGRAHPIC MAP CONSTRUCTION:

To produce a topographic map it is absolutely necessary to determine the elevation of exact locations shown on each map. To obtain this information, two processes can be used ---

       1. Field Surveying,

          and/or

       2. Aerial Photography

FIELD SURVEY:

Construction of topographic maps via field survey involves the determination of elevations across the earth's surface with plane-table instruments. This consists of a flat board covered with paper upon which the elevations can be marked, and an instrument similar to a rifle telescope (called an alidade). A stadia rod is sighted through the alidade to determine exact distances and elevations for plotting on the map of the plane table. After the necessary plottings are made, the map is returned to the laboratory where the map can be drawn (the subject of "cartography"). The use of lasers in surveying is greatly improving both the accuracy and speed of obtaining elevation data.

AERIAL PHOTOGRAPHY:

Most topographic maps are constructed from aerial photographs because they are less expensive (require less time), and more easily mass-produced. This method of map construction requires aircraft flying over the surface in a grid pattern taking thousands of pictures which slightly overlap one another. In the laboratory the use of the "stereoscope" enables these photographs to be read for elevation, distance, and other features necessary for topographic map construction. Field survey is actually more accurate but is much slower and actually more expensive than the aerial photography method. Aerial photography is especially useful in areas of rugged terrain and heavily forested regions. Some sample aerial photographs and a stereoscope are provided for you to see first hand the 3-D images they produce. You can always find the dates of surveying and/or taking of aerial photographs for all topographic maps in the margins of the maps. Check the maps provided for this information and consider the following:

       a. What is a "cultural update"?
       b. Why are some maps rarely updated?
       c. What is a "field check"?

MAP SCALE:

Map scale is the relationship between a unit of distance on a map and the distance it represents for the area in reality. Map scale is expressed in one of three manners:

a. Verbal Scale .... One Inch "Represents" One Mile.

b. Graphic Scale .... 0_____1

                                                  One Mile

c. Representative Fraction 1: 1,000,000

REPRESENTATIVE FRACTION SCALE:

A fractional scale is always given on topographic maps directly above the graphic scale along the bottom margin of the map. This scale is a fraction and refers to the number of inches on the map as compared to the actual distance in inches on the earth's surface. For example, a map scale of 1:24,000 means that each inch on the map represents 24,000 inches on the earth's surface. This is the most important scale on topographic maps! You will need to memorize and be able to use the three most common R.F. Scales: 1:24,000, 1:48,000, and 1:62,500.

To convert the map scale, which is in inches, to miles, you merely change the inches in the scale to feet by dividing by 12, then divide the resulting feet by 5280 (the number of feet in a mile). This will give you the number of miles, or fraction of mile, that each inch represents.

$$\frac{24,000}{12} = \frac{2000}{5280} = .38 \text{ miles} \qquad \frac{62,500*}{12} = \frac{5280}{5280} = 1 \text{ mile}$$

$$\frac{48,000}{12} = \frac{4000}{5280} = .76 \text{ miles}$$

*NOTE: 62,500 is used as the number of inches in 1 mile instead of the actual number of 63,360!

LARGE SCALE VS. SMALL SCALE MAPS:

A. A "large scale" map covers a small area showing good detail. EXAMPLE: A map of Vincennes University campus would be a large scale map. (Figure 20A).

B. A Small scale map covers a large area showing poor detail. EXAMPLE: A map of the world would be a small scale map. (Figure 20B).

C. The larger the fraction, the larger the scale of the map; thus, 1:24,000 is a larger scale than 1:62,500. Remember 1:1 is reality & 1:24,000 is closer to it than 1:62,500.

Campus Map
1:400
Large Scale

Figure 20A.

World Map
1:1,000,000
Small Scale

Figure 20B.

DETERMINING MAP AREA:

Map area can easily be determined by multiplying the length of the map by the width, and using the scale for conversion into the square miles. Area is always in square miles or kilometers so the fractional scale has to be changed from inches into miles or kilometers. To figure map area measure the length of the map in inches and fractions and change that figure into miles and fractions using the scale given: next measure the width of the map in inches and fractions and change that figure into miles and fractions using the scale once again; finally multiply these two answers (length in miles X width in miles) to determine the number of square miles represented by the map.

MAP AREA PROBLEMS:

If the World Map of Figure 20B is 1" square, and has a scale of 1:1,000,000 then it covers _____ sq. miles.

A map with a scale of 1:48,000 that is 10" x 10" covers _____ sq. miles.

ANGLE OF DECLINATION: (View this on topographic maps provided)

The difference in degrees between True North (Geographic North) and Magnetic North (based on compass readings) is referred to as the angle of declination. Magnetic North can be found in the atlas at approximately 74 degrees north and 101 degrees west (along the S. shore of Bathurst Island in Canada). NOTE: The earth's magnetic poles are constantly shifting by small amounts and corrections for map reading are available. Declination on topographic maps is shown on the bottom left corner through the use of the symbol shown below.

WHY IS THE KNOWLEDGE OF DECLINATION SO IMPORTANT TO THE USE OF TOPOGRAPHIC MAPS?

Can any location have a
0° declination?

Example of Magnetic
Declination Symbol

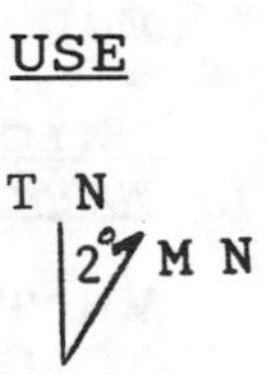

TOPOGRAPHIC MAP SYMBOLS

A number of topographic map symbols are used to show both cultural (man-made), and physical (natural) features on topographic maps. Symbols are easier to use and require less space thus avoiding the creation of a cluttered & difficult to read map. An index sheet of "TOPOGRAPHIC MAP SYMBOLS" is provided on the following page.

How are the following illustrated on a topographic map?

A. an orchard?

B. a church?

C. pipeline?

D. a mine or quarry?

E. a marsh or swamp?

TOPOGRAPHIC MAP COLORS:

The colors used on topographic maps are standardized to always represent the same features: Learn These!

        BROWN .....Contour Lines (elevations)
        BLACK .....Man-made Features (except roads)
        BLUE.......Water Bodies
        GREEN......Vegetation
        RED........Important Roads & Land Survey System
        PURPLE.....Revisions to Map from recent aerial
                   photographs (Not field checked)!

LATITUDE & LONGITUDE COORDINATES:

Latitude & Longitude: The coordinates of latitude & longitude will always be present along the map border. Most topographic maps contain 15' or 7 1/2' of latitude and longitude and are thus referred to as "quadrangeles" (since they cover equal amounts of latitude and longitude on all 4 sides).

Standard edition topographic maps are "Quadrangles" with a R.F. scale of 1:24,000, and cover 7 1/2' of latitude & longitude. Quadrangles with a R.F. scale of 1:62,500 are called 15 minute series maps because they cover 15' of latitude & longitude. The latitude and longitude is given at map corners and at intervals along the sides.

Which is a better map for hiking/hunting - 7 1/2' or 15'? Why?

WHAT OTHER INFORMATION IS PROVIDED IN THE MARGIN OF TOPOGRAPHIC MAPS? WHAT OTHER TYPES OF MAPS ARE AVAILABLE? NOTE: INFORMATION ON HOW TO PURCHASE TOPOGRAPHIC MAPS IS PROVIDED IN THE APPENDIX!

TOPOGRAPHIC MAP SYMBOLS

VARIATIONS WILL BE FOUND ON OLDER MAPS

Primary highway, hard surface . . . .

Secondary highway, hard surface . . . .

Light-duty road, hard or improved surface . . . .

Unimproved road . . . .

Road under construction, alinement known . . . .

Proposed road . . . .

Dual highway, dividing strip 25 feet or less . . . .

Dual highway, dividing strip exceeding 25 feet . . . .

Trail . . . .

Railroad: single track and multiple track . . . .

Railroads in juxtaposition . . . .

Narrow gage: single track and multiple track . . . .

Railroad in street and carline . . . .

Bridge: road and railroad . . . .

Drawbridge: road and railroad . . . .

Footbridge . . . .

Tunnel: road and railroad . . . .

Overpass and underpass . . . .

Small masonry or concrete dam . . . .

Dam with lock . . . .

Dam with road . . . .

Canal with lock . . . .

Buildings (dwelling, place of employment, etc.) . . . .

School, church, and cemetery . . . .

Buildings (barn, warehouse, etc.) . . . .

Power transmission line with located metal tower . . . .

Telephone line, pipeline, etc. (labeled as to type) . . . .

Wells other than water (labeled as to type) . . . . oOil . . . . . oGas

Tanks: oil, water, etc. (labeled only if water) . . . . • ● Water

Located or landmark object; windmill . . . . o . . . .

Open pit, mine, or quarry; prospect . . . . X X

Shaft and tunnel entrance . . . . ▪ . . . . Y

Horizontal and vertical control station:

    Tablet, spirit level elevation . . . . BM △ 5653

    Other recoverable mark, spirit level elevation . . . . △ 5455

Horizontal control station: tablet, vertical angle elevation VABM △95I9

    Any recoverable mark, vertical angle or checked elevation △3775

Vertical control station: tablet, spirit level elevation . . . . . BM ✕ 957

    Other recoverable mark, spirit level elevation . . . . ✕ 954

Spot elevation . . . . ✕ 7369   ✕ 7369

Water elevation 670   670

Boundaries: National . . . .

    State . . . .

    County, parish, municipio . . . .

    Civil township, precinct, town, barrio . . . .

    Incorporated city, village, town, hamlet . . . .

    Reservation, National or State . . . .

    Small park, cemetery, airport, etc. . . . .

    Land grant . . . .

Township or range line, United States land survey . . . . . .

Township or range line, approximate location . . . .

Section line, United States land survey . . . .

Section line, approximate location . . . .

Township line, not United States land survey . . . .

Section line, not United States land survey . . . .

Found corner: section and closing . . . .

Boundary monument: land grant and other . . . . □ . . . . . □

Fence or field line . . . .

Index contour . . . .   Intermediate contour . . .

Supplementary contour . . . . . .   Depression contours . . . . .

Fill . . . .   Cut . . . .

Levee . . . .   Levee with road . . . . .

Mine dump . . . .   Wash . . . .

Tailings . . . .   Tailings pond . . . .

Shifting sand or dunes . .   Intricate surface . . . . .

Sand area . . . .   Gravel beach . . . .

Perennial streams . . . .   Intermittent streams . . .

Elevated aqueduct . . . .   Aqueduct tunnel . . . . .

Water well and spring . .   Glacier . . . .

Small rapids . . . .   Small falls . . . .

Large rapids . . . .   Large falls . . . .

Intermittent lake . . . . .   Dry lake bed . . . .

Foreshore flat . . . .   Rock or coral reef . . . .

Sounding, depth curve . . 10   Piling or dolphin . . . . .

Exposed wreck . . . .   Sunken wreck . . . .

Rock, bare or awash; dangerous to navigation . . . .

Marsh (swamp) . . . .   Submerged marsh . . . .

Wooded marsh . . . .   Mangrove . . . .

Woods or brushwood . .   Orchard . . . .

Vineyard . . . .   Scrub . . . .

Land subject to
controlled inundation   Urban area . . . .

MILE SCALE 1:62 500

**UNITED STATES
DEPARTMENT OF THE INTERIOR
GEOLOGICAL SURVEY**

**TOPOGRAPHIC
MAP INFORMATION AND SYMBOLS
MARCH 1978**

QUADRANGLE MAPS AND SERIES

Quadrangle maps cover four-sided areas bounded by parallels of latitude and meridians of longitude. Quadrangle size is given in minutes or degrees.

Map series are groups of maps that conform to established specifications for size, scale, content, and other elements.

Map scale is the relationship between distance on a map and the corresponding distance on the ground.

Map scale is expressed as a numerical ratio and shown graphically by bar scales marked in feet, miles, and kilometers.

NATIONAL TOPOGRAPHIC MAPS

Series	Scale	1 inch represents	1 centimeter represents	Standard quadrangle size (latitude-longitude)	Quadrangle area (square miles)
7½-minute	1:24,000	2,000 feet	240 meters	7½ × 7½ min.	49 to 70
7½ × 15-minute	1:25,000	about 2,083 feet	250 meters	7½ × 15 min.	98 to 140
Puerto Rico 7½-minute	1:20,000	about 1,667 feet	200 meters	7½ × 7½ min.	71
15-minute	1:62,500	nearly 1 mile	625 meters	15 × 15 min.	197 to 282
Alaska 1:63,360	1:63,360	1 mile	nearly 634 meters	15 × 20 to 36 min.	207 to 281
Intermediate	1:100,000	nearly 1.6 miles	1 kilometer	30 × 60 min.	1568 to 2240
U. S. 1:250,000	1:250,000	nearly 4 miles	2.5 kilometers	1° × 2° or 3°	4,580 to 8,669
U. S. 1:1,000,000	1:1,000,000	nearly 16 miles	10 kilometers	4° × 6°	73,734 to 102,759
Antarctica 1:250,000	1:250,000	nearly 4 miles	2.5 kilometers	1° × 3° to 15°	4,089 to 8,336
Antarctica 1:500,000	1:500,000	nearly 8 miles	5 kilometers	2° × 7½°	28,174 to 30,462

CONTOUR LINES SHOW LAND SHAPES AND ELEVATION

The shape of the land, portrayed by contours, is the distinctive characteristic of topographic maps.

Contours are imaginary lines following the ground surface at a constant elevation above or below sea level.

Contour interval is the elevation difference represented by adjacent contour lines on maps.

Contour intervals depend on ground slope and map scale. Small contour intervals are used for flat areas; larger intervals are used for mountainous terrain.

Supplementary dotted contours, at less than the regular interval, are used in selected flat areas.

Index contours are heavier than others and most have elevation figures.

Relief shading, an overprint giving a three-dimensional impression, is used on selected maps.

Orthophotomaps, which depict terrain and other map features by color-enhanced photographic images, are available for selected areas.

COLORS DISTINGUISH KINDS OF MAP FEATURES

Black is used for manmade or cultural features, such as roads, buildings, names, and boundaries.

Blue is used for water or hydrographic features, such as lakes, rivers, canals, glaciers, and swamps.

Brown is used for relief or hypsographic features—land shapes portrayed by contour lines.

Green is used for woodland cover, with patterns to show scrub, vineyards, or orchards.

Red emphasizes important roads and is used to show public land subdivision lines, land grants, and fence and field lines.

Red tint indicates urban areas, in which only landmark buildings are shown.

Purple is used to show office revision from aerial photographs. The changes are not field checked.

INDEXES SHOW PUBLISHED TOPOGRAPHIC MAPS

Indexes for each State, Puerto Rico and the Virgin Islands of the United States, Guam, American Samoa, and Antarctica show available published maps. Index maps show quadrangle location, name, and survey date. Listed also are special maps and sheets, with prices, map dealers, Federal distribution centers, and map reference libraries, and instructions for ordering maps. Indexes and a booklet describing topographic maps are available free on request.

HOW MAPS CAN BE OBTAINED

Mail orders for maps of areas east of the Mississippi River, including Minnesota, Puerto Rico, the Virgin Islands of the United States, and Antarctica should be addressed to the Branch of Distribution, U. S. Geological Survey, 1200 South Eads Street, Arlington, Virginia 22202. Maps of areas west of the Mississippi River, including Alaska, Hawaii, Louisiana, American Samoa, and Guam should be ordered from the Branch of Distribution, U. S. Geological Survey, Box 25286, Federal Center, Denver, Colorado 80225. A single order combining both eastern and western maps may be placed with either office. Residents of Alaska may order Alaska maps or an index for Alaska from the Distribution Section, U. S. Geological Survey, Federal Building-Box 12, 101 Twelfth Avenue, Fairbanks, Alaska 99701. Order by map name, State, and series. On an order amounting to $300 or more at the list price, a 30-percent discount is allowed. No other discount is applicable. Prepayment is required and must accompany each order. Payment may be made by money order or check payable to the U. S. Geological Survey. Your ZIP code is required.

Sales counters are maintained in the following U. S. Geological Survey offices, where maps of the area may be purchased in person: 1200 South Eads Street, Arlington, Va.; Room 1028, General Services Administration Building, 19th & F Streets NW, Washington, D. C.; 1400 Independence Road, Rolla, Mo.; 345 Middlefield Road, Menlo Park, Calif.; Room 7638, Federal Building, 300 North Los Angeles Street, Los Angeles, Calif.; Room 504, Custom House, 555 Battery Street, San Francisco, Calif.; Building 41, Federal Center, Denver, Colo.; Room 1012, Federal Building, 1961 Stout Street, Denver Colo.; Room 1C45, Federal Building, 1100 Commerce Street, Dallas, Texas; Room 8105, Federal Building, 125 South State Street, Salt Lake City, Utah; Room 1C402, National Center, 12201 Sunrise Valley Drive, Reston, Va.; Room 678, U. S. Court House, West 920 Riverside Avenue, Spokane, Wash.; Room 108, Skyline Building, 508 Second Avenue, Anchorage, Alaska; and Federal Building, 101 Twelfth Avenue, Fairbanks, Alaska.

Commercial dealers sell U. S. Geological Survey maps at their own prices. Names and addresses of dealers are listed in each State index.

INTERIOR—GEOLOGICAL SURVEY, RESTON, VIRGINIA—1978

EXERCISE 5-A
MAP READING

Student Name _____ Lab. Sec. # _____

1. Datum Plane for most topographic maps

    is _____ _____ _____ .

2. What is the major draw back to using field
    survey to prepare topographic maps?

    _____

    _____

3. A 10" x 20" map of Indiana has a
    larger scale than a 10" x 20" map of
    Knox County.   TRUE OR FALSE

4. Indiana has less magnetic declination
    than California.   TRUE OR FALSE

5. A map has a R.F. scale of 1:24,000; a
    distance of 15" represents _____
    miles.

6. What is the symbol for a school on a
    topographic map?   _____

7. Brown is a color used on topographic
    maps exclusively to show _____.

8. A topographic map is 24" x 18" with a
    scale of 1:62,500; what is the area
    covered by the map?   _____ sq. miles.

Using the "Oolitic Quadrangle" provided:

9. What is the latitude & longitude of
    the exact center of the map? (Circle
    N or S; E or W)

    _____ N-S and _____ E-W

10. This is a _____ minute series map; it was surveyed

    in _____; the magnetic declination is _____;

    the topographic map immediately to the SE of this

    map is named the _____ Quadrangle.

123

EXERCISE 5-B
MAP READING

Student Name _____ Lab. Sec. # _____

1. Topographic maps show the "relief" of an area; what does that mean?

   _____

2. Name 2 advantages to the use of aerial photographs in construction of topographic maps.

   a. _____ b._____

3. The closer to a R.F. scale of 1:1 a map gets the larger the scale becomes.   TRUE OR FALSE

4. The earth's magnetic poles are fixed and do not move.   TRUE OR FALSE

5. A map has a R.F. scale of 1:48,000; a distance of 7 1/2" represents _____miles.

6. What is the symbol for a power transmission line on a topographic map?   _____

7. The color purple on a topographic map

   indicates _____.

8. A topographic map is 15" x 20" with a scale of 1:24,000; what is the area covered by the map?
                                                                              _____
                                                                              sq. miles.

Using the "Princeton Quadrangle" provided:

9. What is the latitude & longitude of the exact center of the map? (Circle N or S; E or W)

   _____ N-S and _____ E-W

10. This is a _____ minute series map; it was compiled

    in 1962 from maps surveyed in the year _____; the

    topographic map immediately to the NW of this map is the

    _____ Quadrangle; traveling 6" across

    this map is a distance of _____ miles.

EXERCISE 5-C
MAP READING

Student Name _____ Lab. Sec. # _____

1. Both planimetric & topographic maps
    show cultural features; only
    topographic show relief.   TRUE OR FALSE

2. Small scale maps are more useful to
    compare the size & shapes of world
    countries; large scale maps are
    better for land development studies
    (i.e. malls, airports, etc.)   TRUE OR FALSE

3. A small scale map is more likely to
    have a R.F. scale of 1:10,000 than
    1:100,000.   TRUE OR FALSE

4. Which is more accurate - elevations
    determined from field surveying or
    aerial photography?   _____

5. What is the symbol for a railroad
    tunnel on a topographic map?   _____

6. What color is used on topographic
    maps to show urban (city) areas?   _____

7. A topographic map is 10" x 5" with a
    scale of 1:20,000; what is the area
    covered by the map?   _____
                                                   sq. miles

8. Where is the earth's north magnetic
    pole located?   _____

Using the "Vincennes Quadrangle" provided:

9. What is the latitude & longitude of the
    exact center of the map? (Circle N or S,
    E or W)
    _____ N-S and _____ E-W

10. This is a _____ minute series map; it was last

    surveyed in _____ and updated from aerial photographs

    in _____; the magnetic declination for the map is ____;

    the topographic map immediately to the south of this map

    is the _____ Quadrangle.

127

EXERCISE 5-D
MAP READING

Student Name _____ Lab. Sec. # _____

1. The biggest difference in planimetric and topographic maps is scale.
   TRUE OR FALSE

2. It is possible for the magnetic north pole to be "due north" geographically of some locations.
   TRUE OR FALSE

3. Maps of the world are larger scale than maps of the United States.
   TRUE OR FALSE

4. A stereoscope is an instrument used to

   read _____ _____.

5. What type of map scale does this represent? 0 _____ 10 miles
   _____

6. What is the symbol for a multiple tract railroad on a topographic map?
   _____

7. The use of the color blue on a topographic map indicates

   _____; green indicates _____.

8. A topographic map is 20" x 20" with a scale of 1:250,000; what is the area covered by the map?
   _____
   sq. miles.

Using the "Leavenworth Quadrangle" provided:

9. What is the latitude & longitude of the exact center of the map? (Circle N or S; E or W)

   _____ N-S and _____ E-W

10. This is a _____ minute series map; it was initially

    surveyed in _____ and most recently updated by aerial

    photography in _____; 10" across this map represents

    a distance of _____ miles.

129

EXERCISE 5-E
MAP READING

Student Name _____ Lab. Sec. # _____

1. It is correct to refer to planimetric maps as 2-D and topographic maps as 3-D.   TRUE OR FALSE

2. Aerial photographs are used to check the accuracy of field surveying, not visa versa.   TRUE OR FALSE

3. Maps used for back-packing in the mountains are more likely to be large scale.   TRUE OR FALSE

4. There is more magnetic declination at the top of a topographic map than at the bottom.   TRUE OR FALSE

5. A map has a R.F. scale of 1:50,000; a distance of 8" represents _____ miles.

6. What is the symbol for an intermittant lake on a topographic map?   _____

7. The color black on a topographic map

   is used to represent _____.

8. A topographic map is 25" x 25" with a scale of 1:10,000; what is the area covered by the map?
                                                                        _____
                                                                        sq. miles.

Using the "Bright Angel" Quadrangle provided:

9. What is the latitude & longitude of the exact center of the map? (Circle N or S, E or W)

   _____ N-S and _____ E-W

10. This is a _____ minute series map; it was field

    checked for accuracy in _____; the magnetic

    declination for the map is _____; a trial 10" long

    on this map is _____ miles in length.

EXERCISE 5-F
Map Reading

Student Name _____ Lab. Sec. # ____

1. Topographic maps show relief; relief is the change

   in _____ across the map.

2. If data from field surveying & aerial photographs
   disagree, which is more likely to be correct?

   _____

3. Why are there no maps drawn with a R.F. scale of 1:1?

   _____

4. A map has a R.F. scale of 1:100,000; a distance of 3"

   represents _____ miles.

5. What is the symbol for a bridge on a topographic map?

   _____

6. What is the color brown used to represent on topographic

   maps? _____

7. A topographic map is 8" x 11" with a scale of 1:62,500;
   the map thus covers _____ sq. miles.

8. Walking north parallel to the sides of
   most topographic maps is leading one
   toward the magnetic north pole.   TRUE OR FALSE

Using the "Muncie East Quadrangle" provided:

9. What is the latitude & longitude of the exact center of
   the map? (Circle N or S; E or W)

   _____ N-S and _____ E-W

10. This is a _____ minute series map; it was surveyed in

    _____ and last field checked in _____; Prairie

    Creek is approximately 1 1/2" across thus _____ feet

    from side to side; the topographic map immediately to the

    north of this map is the _____ Quadrangle.

EARTH SCIENCE   LAB. EXERCISE #6

"READING CONTOUR LINES"

- -

OBJECTIVES:

Be Able to:

1. ACCURATELY USE THE BASIC RULES OF CONTOUR
    LINES WHEN READING TOPOGRAPHIC MAPS.

2. CONSTRUCT A CONTOUR MAP GIVEN A SET OF
    KNOWN ELEVATIONS.

3. INTERPRETATE SUCH TOPOGRAPHIC FEATURES
    AS DEPRESSIONS, HILLS, RIVER VALLEYS,
    DEGREE OF SLOPE, ETC., FROM THE CONTOUR
    LINES ON TOPOGRAPHIC MAPS.

4. DETERMINE MAXIMUM AND MINIMUM POSSIBLE
    ELEVATIONS ON TOPOGRAPHIC MAPS.

5. DEFINE AND/OR EXPLAIN THE GLOSSARY TERMS.

"I GUESS THAT BROWN AREA ON THE MAP WASN'T THE BEACH AFTERALL."

GLOSSARY TERMS:

CONTOUR LINE:

CARTOGRAPHY:

CONTOUR INTERVAL:

INDEX CONTOURS:

HACHURED CONTOURS:

RELIEF:

LOGICAL CONTOURING:

INTERPOLATION:

COMMON CONTOUR INTERVALS:

MAXIMUM TOPOGRAPHIC ELEVATION:

MINIMUM TOPOGRAPHIC ELEVATION:

GENERAL TOPOGRAPHIC RELIEF:

READING CONTOUR LINES:

Learning the following applications of contour lines will enable you to "read" topographic maps and determine the amount and location of relief.

Imagine that for each ten feet of elevation above sea level on an island you painted a line around the island connecting points of equal elevation. <u>Figure 21</u> below shows the appearance of such an island. If you then viewed the island from an airplane you would see the lines as represented by <u>Figure 22.</u> This is basically the approach of contour lines to show elevations and locations of physical features on topographic maps.

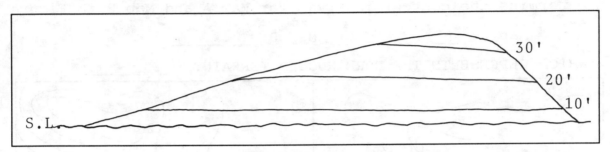

<u>Figure 21.</u>

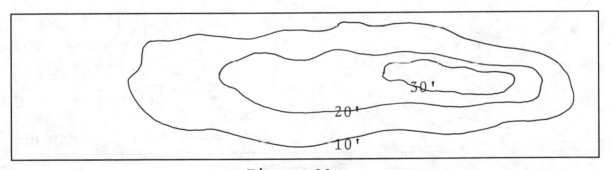

<u>Figure 22.</u>

===
The collection of an adequate quantity of elevations allows the cartographer (map-maker) to draw contour lines which connect the points of equal elevation. Observe the map of Paradise Island below and notice how contour lines could be drawn.

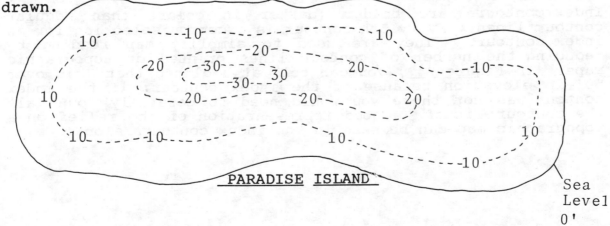

<u>PARADISE ISLAND</u>

Sea Level 0'

CONTOUR INTERVAL:

This refers to the change in elevation, that is, how much relief there is as you move from one contour line to the next. Figure 22 illustrates a contour interval of 10!

The contour interval is given in the marginal information of the map (usually at the bottom). The contour interval will vary from one map to another depending upon the amount of relief. The more rugged the terrain, the larger the contour interval that will be needed. For regions of flat topography, however, a contour interval of only 5 to 10 feet could be used. Commonly used contour intervals include: 5, 10, 20, 50, 100, and 500 feet.

* Determine the contour interval for Map A and Map B in Figure 23.

Map A: _20_   Map B: _50_

WHICH REPRESENTS THE MORE RUGGED TERRAIN? _B_

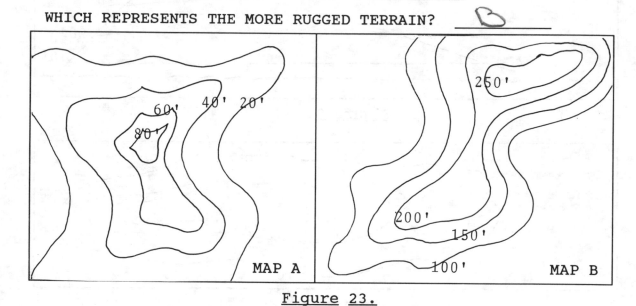

Figure 23.

==

INDEX CONTOURS:

Index contours are bolder (darker in color) than regular contour lines. Every 5th contour is a common selection for an index contour. They are used to simplify map reading by reducing the number of contour lines printed on topographic maps. In Figure 24., you can tell at a glance that X is over 50' in elevation because of the index contour; if the index contour was not there you would need to carefully count all the contour lines! A good representation of the relief on a topographic map can be seen in the index contours alone!

138

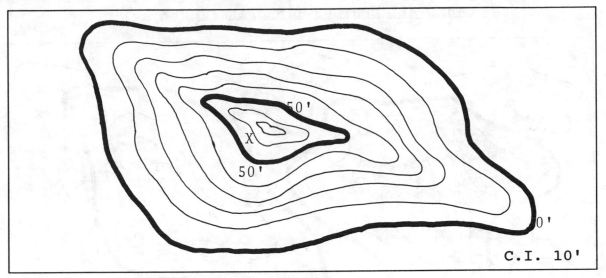

Figure 24.

==

HACHURED CONTOURS:

Special contour lines are used to show depressions on a map
--these are called "hachured" or depression contours and have
small tick marks on the inside to indicate a decrease or drop
in elevation. These tick marks are the hachures and point
toward the bottom of the depression. Hachured contour lines
therefore enclose areas of lower elevation. NOTE: the value
of a hachured contour line is the same value as the adjacent
contour line of lowest value. Study Figure 25.

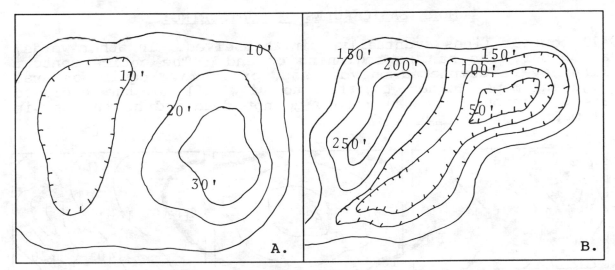

Figure 25.

LABEL ALL CONTOUR LINES IN FIGURE 26.

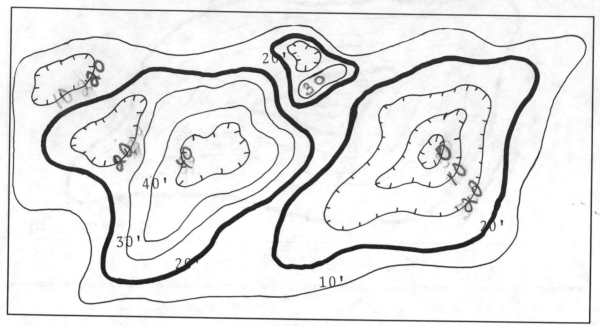

Figure 26.

===

REMEMBER: The elevation of a hachured contour lines is the
same as the closest <u>lower value</u> contour line, <u>**not**</u>
merely the closest contour lines!

===

MORE BASIC RULES OF CONTOURING:

① →All contour lines eventually join themselves. In other words,
a contour line has no beginning or end. The entire contour
line may be completed on your map, or it may require several
adjacent maps joined together to view all of the enclosed
area. Contour lines therefore <u>do</u> <u>not</u> dead-end at the margin
of maps. <u>Study Figure 27.</u>

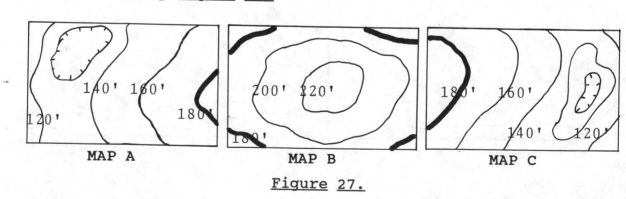

Figure 27.

140

(2) Contour lines __cannot__ cross other contour lines of different value. This is true since no single point can have two different elevations, which is what crossing contour lines would indicate! Contour lines __can__ merge, however, in places where there is great relief. __Figure__ __28.__ illustrates an __impossible__ __situation__! It would be impossible for location X to have the value of both contour lines of different value.

" __AN IMPOSSIBLE SITUATION__ "

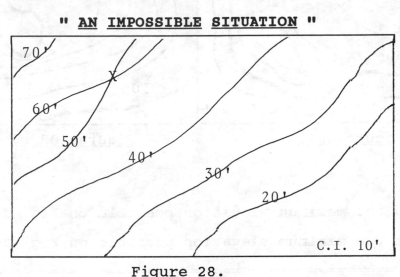

Figure 28.

==

(3) Contour lines always enclose higher ground such as hills or mountains. The exact elevation of the hill or mountain may be written in as a bench mark (see __Figure__ __29.__); otherwise, the elevation of the hill/mountain is expressed as:

   a.   at least as high as the last drawn contour line; but,
   b.   less than the value of the next higher, undrawn
         contour line.

IF BENCH MARK 471' WAS NOT PRESENT, THE HILL COULD BE FROM 460' to 479' HIGH.

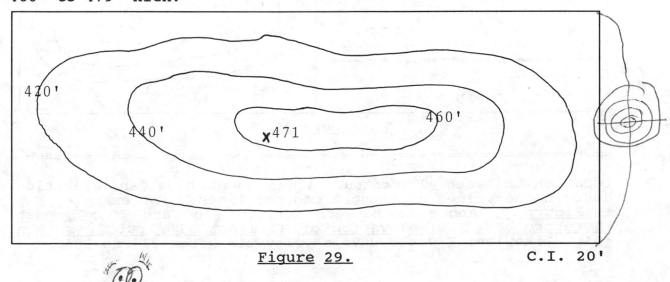

Figure 29.   C.I. 20'

141

MAXIMUM OR MINIMUM RELIEF:

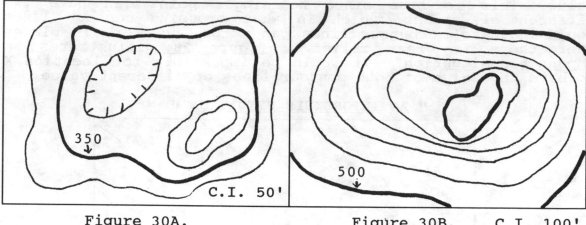

Figure 30A.   Figure 30B.   C.I. 100'

C.I. 50'

PROBLEMS:

1. What is the maximum elevation possible on Fig. 30A? _499_

2. What is the minimum elevation possible on Fig. 30B? _401'_

==

Contour lines have a high and low side; moving far enough in any direction will eventually result in changing to higher or lower elevation. Study Figure 31.

NOTE: X IS ON THE HIGH SIDE AND Y IS ON THE LOW SIDE FOR BOTH!

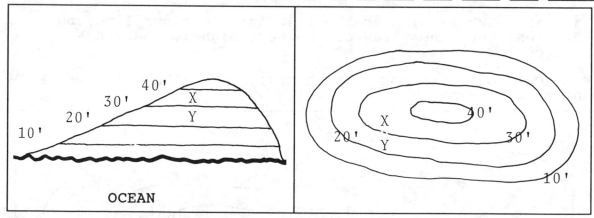

Figure 31.

==

Locations between 2 contour lines must have an elevation between the values of those 2 contour lines. For example, X in Figure 31 above is between contours 30' and 40'; X must therefore be at an elevation of at least 31', but less than 40'. Likewise, Y would have an elevation from 21' to 29'.

142

When contour lines are widely spaced, a low, gentle slope is indicated; when contour lines are close together, a steep slope is indicated. Where contours merge, a cliff or vertical drop is present. Study Figure 32.

Which is a more difficult walk? A to B, or C to B? _____

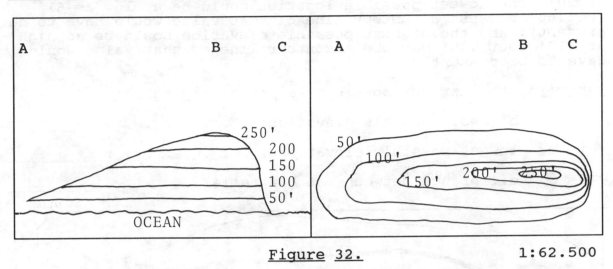

Figure 32.

1:62.500

==

In Figure 32 the distance from A to B is approximately 2 1/4 inches or 2.25 miles, and the elevation change from A to B is at least 250'. The distance from B to C is approximately 1/2 inch or .5 mile, and the elevation change from B to C is the same (at least 250'). It is obvious that A to B is a more gradual slope than B to C. Notice how the contour lines drawn for this island reflect this different degree of slope.

==

Contour lines bend (point) upstream when crossing a river or stream. When a contour line crosses a river valley, the line must bend upstream in order to keep its correct elevation. Since all rivers flow down-slope, the bent contour line can be used to determine the direction of stream flow. As mentioned earlier, the closer together the contours the greater the relief (gradient) of the valley. If the contours form a V-shape, a narrow river valley is present; U-shaped contours would indicate a wider valley. Study Figure 33 to see these relationships.

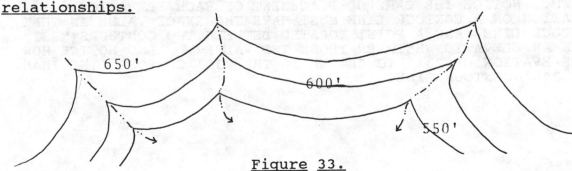

Figure 33.

<u>Relief</u> on a map is the difference between the highest and lowest possible elevations. Even when bench marks are not present you can determine the highest and lowest <u>possible</u> elevations, based on the contour lines. <u>For example</u> : if a lowest value contour line was 80' and the highest value contour line was 460' on a map with a contour interval (C.I.) of 20', the lowest possible location could be as low as 61' (not 60' because a contour line of that value would have to be present), and the highest possible elevation could be as high at 479' (not 480' because a contour line of that value would have to be present).

<u>Determine</u> <u>the</u> <u>maximum</u> <u>possible</u> <u>relief</u> <u>of</u> <u>Figure</u> 34:

    a. highest possible elevation: . . . . <u>549</u>

    b. lowest possible elevation . . . . <u>151</u>

    c. difference between a & b = relief . . . . . <u>398</u>

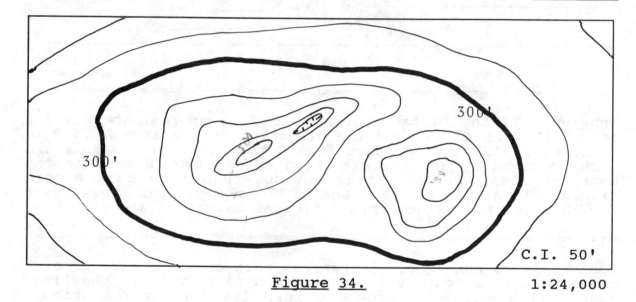

<u>Figure</u> <u>34.</u>   1:24,000

===

BE SURE YOU UNDERSTAND THE RULES FOR CONTOUR LINES! THE DIAGRAM ON THE NEXT PAGE ILLUSTRATES HOW A CONTOUR MAP IS DRAWN. NOTICE THE CAREFUL PLACEMENT OF EACH ELEVATION -- A LOCATION ON A CONTOUR LINE MUST HAVE THE EXACT VALUE AS THAT CONTOUR LINE, AND A POINT LOCATED BETWEEN TWO CONTOURS MUST HAVE AN ELEVATION BETWEEN THOSE TWO VALUES. ALSO NOTICE HOW AN ELEVATION OF 231' IS CLOSER TO THE 230' CONTOUR LINE THAN THE 240' CONTOUR LINE!

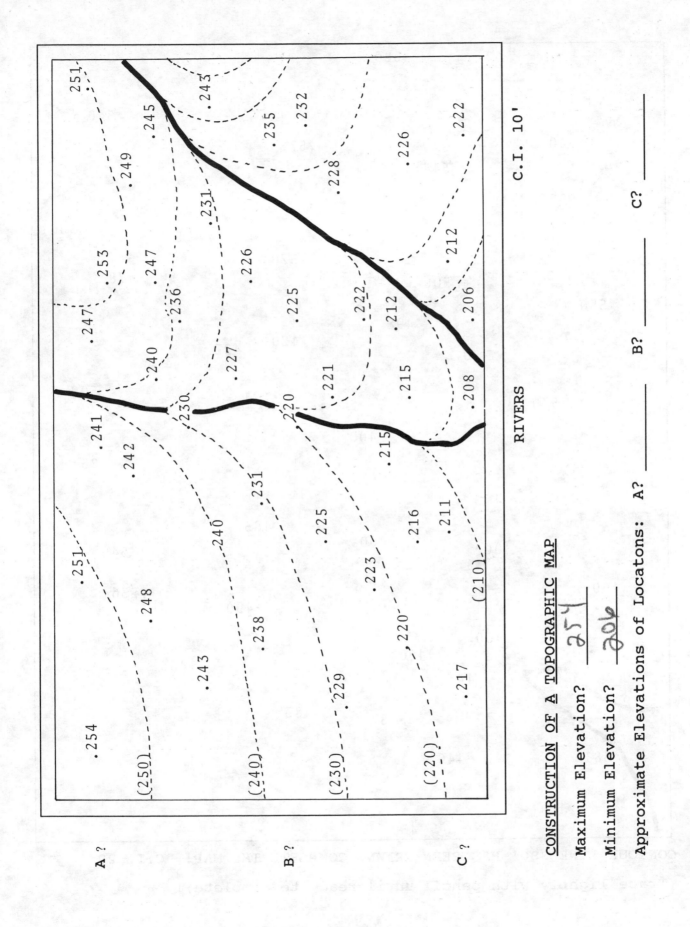

CONSTRUCTION OF A TOPOGRAPHIC MAP

Maximum Elevation? 25⁴

Minimum Elevation? 206

Approximate Elevations of Locatons: A? _____   B? _____   C? _____

C.I. 10'

RIVERS

145

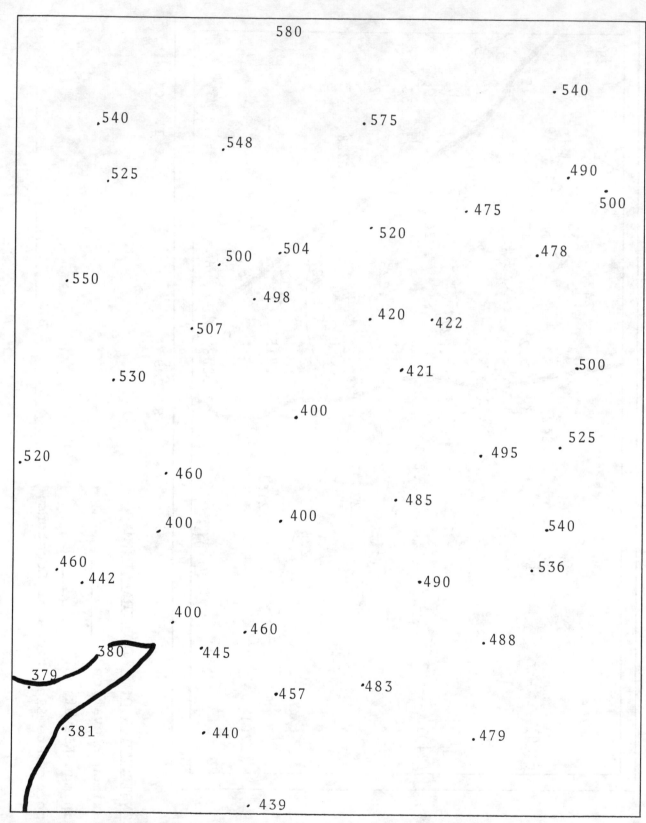

580

.540

.540

.575

.548

.525

.490

.500

.475

.520

.500 .504

.478

.550

.498

.507

.420 .422

.530

.421

.500

.400

.525

.520

.495

.460

.525

.485

.400 .400

.540

.460

.536

.442

.490

.400

.460

.488

.380

.445

.379

.483

.381

.457

.440

.479

.439

CONTOUR LINE 380' HAS BEEN DRAWN. COMPLETE THE MAP!   C.I. 20

(Trace lightly with pencil until ready to complete)!

146

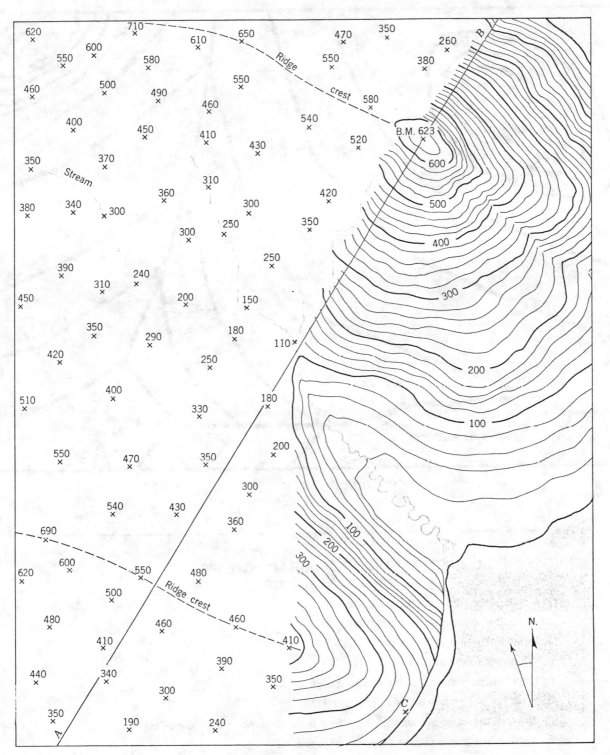

Practice Exercise: "Logical Contouring"

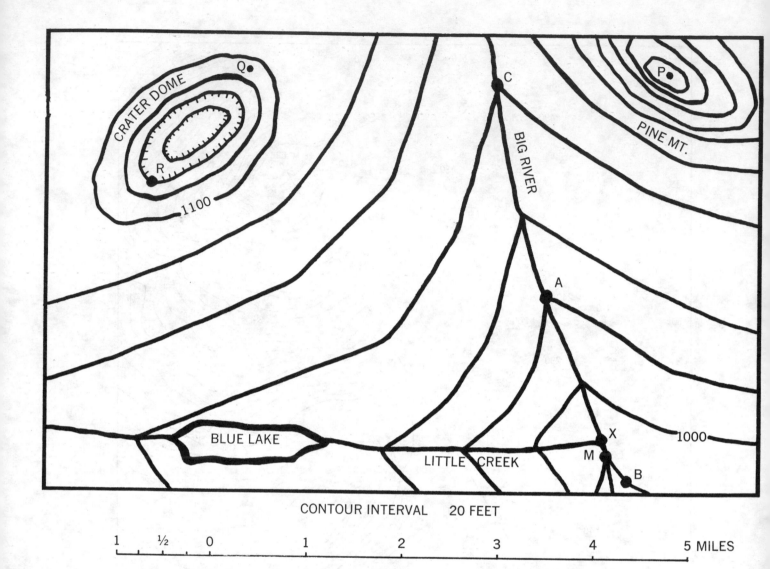

CONTOUR INTERVAL   20 FEET

1   ½   0   1   2   3   4   5 MILES

UNDERSTANDING TOPOGRAPHIC MAPS
CONTOUR MAP EXERCISE

1. What is the elevation of point A?

2. What is the elevation of point B?

3. What is the elevation of point Q?

4. What is the elevation of point R?

5. How much higher is point C than point M?

6. How far is it from point C to point M?

7. What is the average slope of Big River from C to M in feet per mile?

8. What is the approximate elevation of point X?

9. What is the approximate elevation of Blue Lake?

10. What is the highest possible elevation of the top of Pine Mountain?

11. What is the lowest possible elevation of the bottom of the crater of Crater Dome?

12. How far is it, to the nearest quarter mile, from point Q to point P?

13. In which compass direction does Big River flow?

14. In which compass direction does Little Creek flow?

15. Which side of Pine Mountain is steepest?

148

PRACTICE EXERCISE

> 489
397
1'

Using the "Vincennes Quadrangle", answer the following:

1. What series is this map?   *7.5*

2. What is the contour interval for this map?   *10'*

3. What is the map scale for this map?   *1:24000*

4. What is the angle of declination?   *½°*

5. What is the area in sq. miles?

6. What is the highest elevation **possible**?   *489*

7. What is the lowest elevation **possible**?   *398*

8. What is the maximum **possible** relief?   *11"*

9. What is the elevation of Robeson Pond?   *405*

10. What is the elevation of Hutton School on the south edge of Robeson Hills?   *570*

11. What is the **highest possible** elevation for the surface of the Wabash River at the SW border of the map?   *419*

12. What is located at approximately 38°44'00"N. and 87°33'45"W. ?   *Lake Lavergne*

13. Crossing the Lincoln Memorial Bridge from Vincennes, one enters Westport, IL. Northwest of Westport is a water-filled gravel pit. What is its maximum depth?

14. What is the approximate length of Otter Pond?   *¾ of a mile*

15. Where are the steepest slopes located?   *Robeson Hills*

16. Why does the Wabash River flow nearly due south into northern Vincennes?

Because all the steepest hills are located to the north

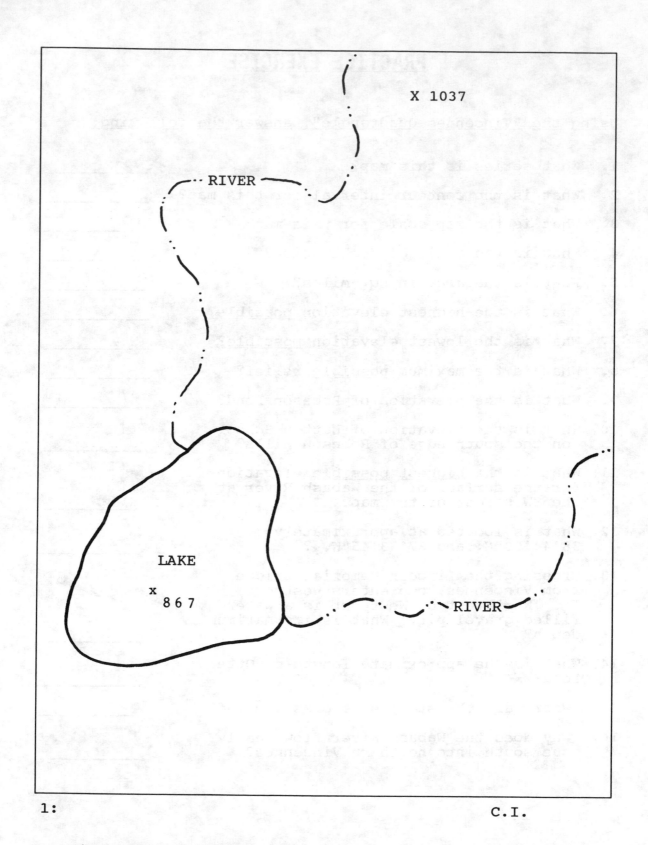

X 1037

RIVER

LAKE

x
8 6 7

RIVER

1:   C.I.

Practice Contour Map Exercise: Create a topographic map
showing all the relationships discussed. Be artistic!

150

EXERCISE 6-A
READING CONTOUR LINES

Student Name: _____ Lab. Sec. # _____

1. Usually every 2nd. contour line of
    different value is an index contour.   TRUE   FALSE

3. Contour lines never cross or merge; such
    situations are impossible.   TRUE   FALSE

4. Contour lines may enclose an area on 1
    topographic map or extend onto the
    adjacent map.   TRUE FALSE

2. Hachured contour lines always indicate

    the presence of a _____.

Use the Bright Angel Quadrangle provided: (5. - 10.)

5. The contour interval is _____; the scale is _____.

6. A 2400' contour lines crosses the Colorado River near

    the center of the map; this is the only contour lines

    that crosses the river on the map. What is the maximum

    possible drop of the river? _____

7. What is the maximum elevation possible for the Grand

    Canyon Lodge in the NE area of the map? _____

8. Compare the Coconino Plateau to the Kaibab Plateau;

    which is most representative? (Grand Canyon is in middle)

    C \/ K   C \/ K   C \/ K

10. What is most likely the maximum elevation
    on Shiva Temple in the N-Central region?   _____

9. There are 100's of depressions shown
    on this map.   TRUE   FALSE

EXERCISE 6-B
READING CONTOUR LINES

Student Name: _____ Lab. Sec. # _____

1. As the amount of relief decreases, the contour interval will increase.   TRUE   FALSE

2. Hachured contour lines have the same value as the adjacent contour line of _____ value.

3. The closer together that contour lines occur, the _____ the degree of slope.

4. Study of topography refers to the earth's _____.

<u>5.</u> - <u>10.</u>   <u>Use</u> <u>the</u> <u>Princeton</u> <u>Quadrangle</u> <u>provided:</u>

5. The contour interval is _____; the scale is _____.

6. No contour lines cross the Wabash River on this map; the <u>maximum</u> drop of the river therefore could not exceed _____.

7. What is the maximum elevation possible for the St. Lucas Church in the SE corner of the map? _____

8. Which has greater relief (steeper slopes) Orrville Hills <u>or</u> Gordon Hills?   _____

9. Is there evidence in any contour lines that the construction of railroads alters surface relief?   YES   OR   NO

10. What is most likely the maximum elevation of Sand Hill in the W-Central map area?   _____

153

EXERCISE 6-C
READING CONTOUR LINES

Student Name: _____ Lab. Sec. # _____

1. It is possible to determine the
   direction of flow of rivers if contour
   lines cross them.   TRUE   FALSE

2. It is possible for a series of hachured
   contour lines to occur within an enclosed
   area. i.e.   TRUE   FALSE

3. Relief on a topographic map is the
   difference between the highest & lowest
   contour lines values; it <u>cannot</u> be more.   TRUE   FALSE

4. What is the appearance of "index contours"
   on a topographic map?

   _____

<u>5. - 10.</u>   Use the Leavenworth Quadrangle provided:

5. The contour interval is _____; the scale is _____.

6. Two contours line crosses the Blue River on this
   map; the maximum drop in elevation possible is therefore
   _____.

7. What is the maximum elevation possible for the picnic
   area in Section 2 in the E-Central map area?   _____

8. What is most unusual about the contours in the Indiana
   vs. Kentucky portions of this map?

   _____

9. As you travel <u>downstream</u> on the Ohio River,
   there are more flat areas (flood plains)
   on which side of the river?   RIGHT   LEFT

10. Why are there very few cultural features on this map?

   _____

155

EXERCISE 6-D
READING CONTOUR LINES

Student Name: _____ Lab. Sec. # _____

1. Only elevations above sea level are
    shown on topographic maps; there
    cannot be negative contours.   TRUE   FALSE

2. It is <u>not</u> possible for hachured contours
    to occur on hills or mountains.   TRUE   FALSE

3. As the amount of known elevations for
    a map increases, the accuracy of the
    contour lines increases.   TRUE   FALSE

4. A number such as x 837 found between
    the 800' and 850' contours represents
    a _____ _____.

<u>5.</u> - <u>10.</u> Use the <u>Vincennes</u> <u>Quadrangle</u> <u>provided:</u>

5. The contour interval is _____; the scale is _____.

6. A 400' contour line is the only contour line to cross
    the Wabash River on this map; the maximum possible drop
    in elevation of the river on this map is therefore
    _____.

7. What is most likely the maximum elevation possible in
    the Robeson Hills? _____

8. There is no portion of the Wabash River
    on this map that does not have levees
    along both sides. (Use symbols sheet)   TRUE   FALSE

9. What appears to be the origin for many
    of the small lakes on the map?   _____

10. Why have there been so many ditches dug
    across the Illinois portion of the map?

    _____

    _____

157

EXERCISE 6-E
READING CONTOUR LINES

Student Name: _____   Lab. Sec. # ____

1. The contour interval for most topo-
    graphic maps is constant and unchanging
    everywhere on that map.   TRUE   FALSE

2. As one crosses hachured contour lines,
    elevations drop at the same contour
    interval as regular contours.   TRUE   FALSE

3. As a contour line crosses a river, the
    line will bend downstream.   TRUE FALSE

4. A contour line has a value of 1250'; 0' on this

    map is the _____ _____. (Not sea level)

5. - 10.   Use the Cumberland Quadrangle provided:

5. The contour interval is _____; the scale is _____.

6. Where is the greatest relief (steepest slope) located

    on this map? _____

7. What is the maximum possible elevation on Haystack

    Mountain in the S-Central area of the map?   _____

8. Which general direction does the North Branch of the
    Potomac River flow?
                            Left to Right or Right to Left

9. What is the most likely natural danger

    facing the majority of people on this

    map?   _____

10. Relief on this map exceeds 1800'.   TRUE   FALSE

EXERCISE 6-F
READING CONTOUR LINES

Student Name: _____   Lab. Sec. # _____

1. A contour interval is more likely to
    be 20' or 50' than 17' or 53'.   TRUE   FALSE

2. The <u>surface</u> <u>level</u> of all lakes will
    be shown on topographic maps as a
    hachured contour.   TRUE   FALSE

3. If the highest value contour line on
    a topographic map is 1250', and the
    contour interval is 50, the highest
    elevation possible is _____.

4. If contour lines merge a _____ is present.

5. - 10.   Use <u>the</u> <u>Oolitic</u> <u>Quadrangle</u> <u>provided:</u>

5. The contour interval is _____; the scale is _____.

6. A 460' contour line crosses the E. Fork of the White
    River near Williams; this is the only contour line that
    crosses the river on the map. What is the maximum
    possible drop of the river?   _____

7. What is the maximum elevation possible on
    the top of Rariden Hill in the SE corner of
    the map?   _____

8. What do the contour lines indicate about
    the relief on the outside bends of the
    White River?

    _____

9. There are many (over 50) depressions on
    this map.   TRUE   FALSE

10. What is a typical elevation within the
    town of East Oolitic?   _____

161

EARTH SCIENCE   LAB. EXERCISE #7

"STREAM GRADIENT & TOPOGRAPHIC PROFILE"

- -

OBJECTIVES:

Be Able to:

1. EXPLAIN AND DETERMINE STREAM GRADIENT.

2. EXPLAIN AND CONSTRUCT TOPOGRAPHIC PROFILES.

3. DETERMINE THE VERTICAL EXAGGERATION OF
    TOPOGRAPHIC PROFILES.

4. APPLY THE RULES OF CONTOURING TO STREAM
    GRADIENT DETERMINATION & PROFILE CONSTRUCTION.

5. DEFINE AND/OR EXPLAIN THE GLOSSARY TERMS.

GLOSSARY TERMS:

Stream Gradient:

Average Vs. Actural Gradient:

V'Ing Contours:

Maximum Stream Gradient:

Minimum Stream Gradient:

Maximum Possible Elevations:

Minimum Possible Elevations:

Topographic Profile:

Logical Profiling:

Profile Line:

Vertical Exaggeration:

Horizontal Scale:

Vertical Scale:

EXERCISE #7

STREAM GRADIENT & PROFILE

Necessary to the satisfactory completion of this exercise is
an understanding of the information covered in exercises 5 &
6. In this exercise you will learn to figure stream gradient,
and how to construct a topographic profile. This skill will
enable you to determine the cross-sectional profile appearance
of any area on a topographic map.

DETERMINATION OF STREAM GRADIENT:

Stream gradient refers to the drop in elevation of a stream
over a given distance. All streams have a gradient or they
would not flow! Could a stream ever flow below sea level?
The answer is yes as long as it is draining to a point that is
lower than sea level (i.e. Dead Sea). Such areas are known as
interior drainage basins.

A typical stream gradient might be expressed in the following
manner: 12.5'/mile. This means that the stream drops an
average of 12.5'/mile over that portion of the course for
which gradient was determined. It does not mean that the
stream slopes uniformly however for it may in reality have
little drop at first, and then begin to drop very fast later
on in its course! Stream gradient is merely an average and
does not indicate the actual gradient of the river for the
entire length of flow!

Stream Gradient Formula: Gradient = $\dfrac{\text{Drop in feet}}{\text{Distance in miles}}$

To determine stream gradient involves two functions:

1.)   Using contour lines, determine the difference
      in elevation between 2 locations on the stream.
      Remember that the contour lines bend upstream
      thus the water flow will be in the opposite
      direction of the V'ing contours. Elevation
      difference can easily be calculated by reading
      the contour lines involved. Study Figure 35.
      Note that the stream is flowing from A toward B.

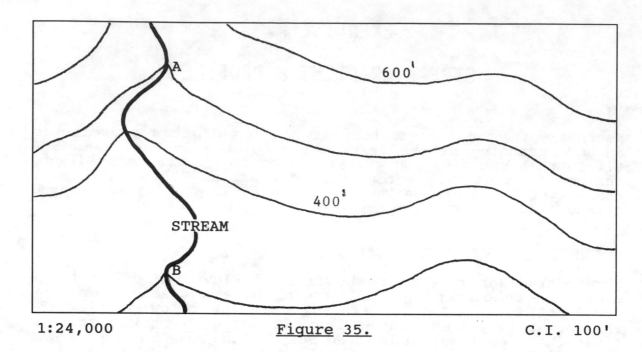

1:24,000   Figure 35.   C.I. 100'

===

2.)   Measure the distance separating the two locations
      on the map. To accurately determine distance a string
      is used. Use of a string allows you to measure dis-
      tance along the stream which is not in a straight
      line. Once the total distance has been measured,
      a ruler is used to convert the string distance into
      feet or miles. This will require use of the map scale;
      for example, if 2 locations are 2.5 inches apart on
      a map with a scale of 1:48,000, they are 1.9 miles
      apart (.76 miles x 2.5)?

      Once you have determined the difference in elevation
      between two locations plus the distances separating
      them, you merely divide the difference in elevation
      by the difference in distance. Consider the
      following:

      If a topographic map had a scale of 1:24,000 and
      contour lines crossed a stream at 480 and 380 feet
      over a distance of 10 inches, what would the
      stream gradient be?

        1. Difference in elevation = 100 feet.
        2. Scale = 1:24,000 so 1" = .38 mile
        3. Distance apart in inches (10) x .38 mi. = 3.8 mi.
        4. Divide the elevation by the distance:

           $$\frac{100'}{3.8 \text{ mi.}} = 26.3'/\text{mile} \underline{\text{(stream gradient)}}$$

166

DETERMINE THE STREAM GRADIENTS FOR FIGURE 36.

S.G. from Location 1-2 $\underline{26.3 \frac{FT}{mile}}$ S.G. from Location 2-3 $\underline{70.2 \frac{FT}{mile}}$

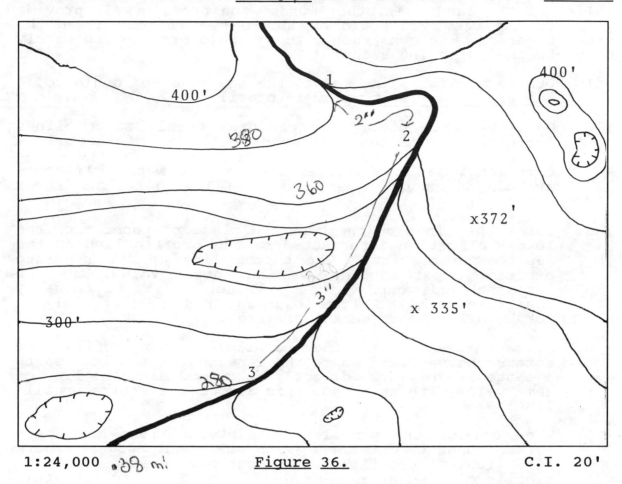

1:24,000 .38 mi.   Figure 36.   C.I. 20'

===

WORK HERE:

$$\frac{380-360}{2'' \times .38} = \frac{20}{.76}$$

$$\frac{360-280}{3'' \times .38}$$

$\frac{.38}{2}$

$\frac{80}{1.14}$

$1.14\overline{)8000}$ 70.2 $\frac{FT}{mile}$ $1.14\overline{)3000}$

$1.14\overline{)}$

1-3 380-

CONSTRUCTION OF THE TOPOGRAPHIC PROFILE:

A topographic profile is simply the cross-section view of the relief of the land. In other words, the profile will provide you with the view you would expect to see if you viewed the area in person. To construct a topographic profile you merely complete the following steps:

1. Draw a straight line across the map at a 90° angle to the area you wish to view in profile;

2. On a separate piece of paper draw evenly spaced lines for each of the contour lines that was intersected by the profile line drawn across the map (make these the same length as the profile line on the map). If there are many contour lines, you can use only the index contours!

3. Label the lines on the separate piece of paper from the lowest elevation intersected by the profile line on the map to the highest. The bottom line on the separate piece of paper should have the lowest value, the top line the additional highest value. (It is actually desirable to have 1 line lower and higher, that will not be used, to frame the profile).

4. Place the series of lines parallel to the profile line across the map, and where the line on the map intersects a contour line, drop directly below and mark the line of same value with an x. Do this across the entire profile line!

5. When all of the profile line intersections have been marked connect the marks with one continuous, smooth line (logical profiling). You can now see the general profile of the area. Study Figure 37. to see this procedure.

VERTICAL EXAGGERATION: (Represented by the letters V.E.)

Nearly all profiles are exaggerated; that is, they give the indication of greater relief than actually exists. To determine how much a profile is exaggerated you merely divide the horizontal map scale for the profile by the vertical scale. The horizontal scale will be the same as that of the topographic map; you can determine the vertical scale by seeing how much elevation is covered in one inch of the series of parallel lines you have drawn. Vertical exaggeration can therefore be controlled by changing the space between the lines of the profile. Normally, the lower the vertical exaggeration the better.

$$\text{Vertical Exaggeration} = \frac{\text{Horizontal Scale}}{\text{Vertical Scale}} \quad \text{(BOTH IN FEET)}$$

SEE APPLICATION OF VERTICAL EXAGGERATION IN FIGURE 37.

CONSTRUCTION OF A TOPOGRAPHIC PROFILE

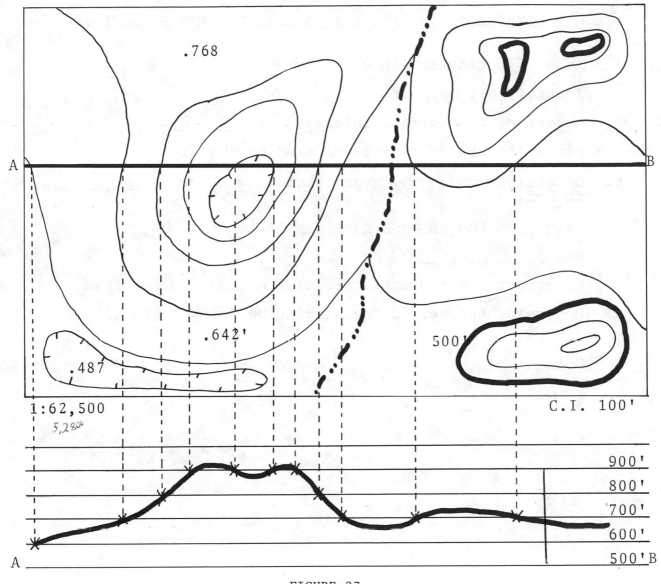

FIGURE 37.

VERTICAL EXAGGERATION: $\dfrac{\text{(horizontal scale)}}{\text{(vertical scale)}} = \dfrac{5280'}{400'} = 13.2x$

comes from 1"

What would the V.E. have been if the C.I. had been 500'? _____

169

PRACTICE EXERCISE

USE MAP LIMBERLOST ON PAGE 171 TO COMPLETE THE FOLLOWING:

1. Label all contour lines;

2. Identify all symbols;

3. Determine the contour interval ___20'___ ;

4. Determine the total area covered by the map _____ ;

5. Determine the stream gradient from A to B _____ ft./mi. ;

6. Determine the maximum possible relief for the map _____ ;

7. Construct a topographic profile for the profile line;

8. Determine the vertical exaggeration for the profile.

WORK HERE:

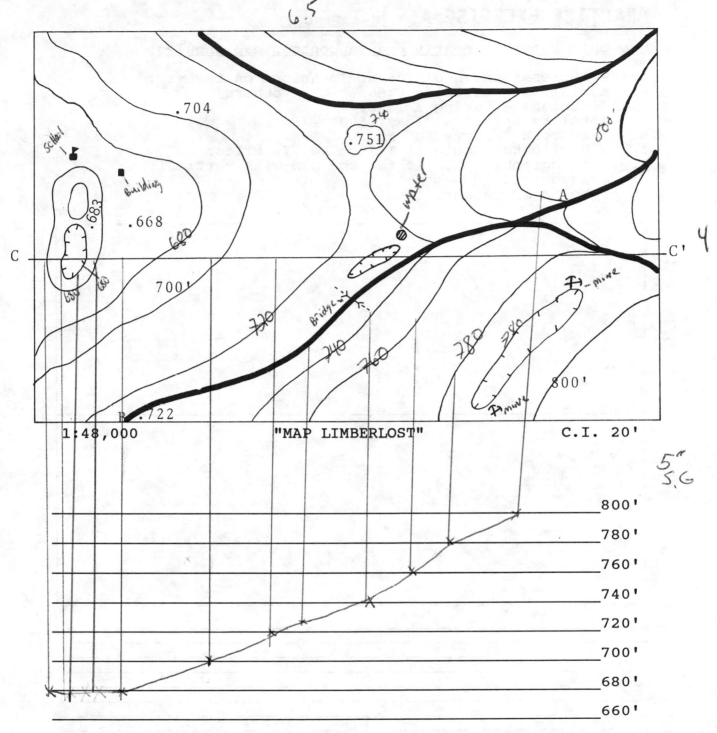

1:48,000   "MAP LIMBERLOST"   C.I. 20'

MAP LIMBERLOST PROFILE

PRACTICE EXERCISE A: (Use Pencil)

Use this sheet to neatly draw a contour map showing:

a. 2 depressions in different locations on the map;
b. a hill with one side steeper than others;
c. 2 streams entering a lake;
d. total relief of at least 600';
e. a contour interval of 50';
f. a horizontal scale of 1 inch to .76 miles;
g. a topographic profile for the map with vertical exaggeration;
h. label all contour lines!

VERTICAL EXAGGERATION? _____   Student Name _____

PRACTICE EXERCISE B: (Use Pencil)

Use this sheet to neatly draw a contour map showing:

a. a depression with 3 interior contour lines;
b. a hill with a bench mark elevation;
c. a stream that drops at least 800 feet;
d. total relief of at least 1000 feet;
e. a contour interval of 100 feet;
f. a horizontal scale of 1 inch to .38 miles;
g. a topographic profile for the map with vertical exaggeration;
h. label all contour lines!

VERTICAL EXAGGERATION? _____   Student Name _____

EXERCISE 7-A
STREAM GRADIENT & TOPOGRAPHIC PROFILE

Student Name: _____   Lab. Sec. # _____

1. Contour lines bend in the direction
    streams flow as they cross the channel.   TRUE   FALSE

2. Stream gradient is the average and not
    the actual drop in elevation from one
    stream location to another.   TRUE   FALSE

3. By changing the vertical scale on a topo-
    graphic profile the V.E. can be controlled.   TRUE   FALSE

4. It may not be necessary to use all con-
    tour lines when constructing a topo-
    graphic profile.   TRUE   FALSE

5. Vertical exaggeration of 100 x is always
    better than a V.E. of 10 x.   TRUE   FALSE

6. A topographic profile can be constructed
    for only certain areas on a topographic
    map.   TRUE   FALSE

7. - 10. Draw a profile for the map below:

1:50,000   C.I. 50'   V.E. = _____

175

EXERCISE 7-B
STREAM GRADIENT & TOPOGRAPHIC PROFILE

Student Name: _____   Lab. Sec. # _____

1. Stream gradient decreases as the
    relief on the map increases.   TRUE   FALSE

2. Horizontal scale must be changed
    into feet before vertical exaggeration
    can be figured.   TRUE   FALSE

3. If a topographic map had a contour inter-
    val of 20 ' with the highest contour line
    960' and the lowest contour line 720',
    total relief on the map <u>could</u> <u>not</u> be as
    much as 278'.   TRUE   FALSE

4. Contour lines cross a stream at 350' and
    200' over a distance of 25 miles. What
    is the stream gradient? (scale = 1:24,000)   _____

5. It is possible to control the vertical
    exaggeration on topographic profiles.   TRUE   FALSE

6. In drawing a topographic profile using
    only index contours, an entirely dif-
    ferent profile with different elevations
    and slopes would result.   TRUE   FALSE

7. - 10. Draw a profile for the map below:

1:48,000   C.I. 50'   V.E. = _____

EXERCISE 7-C
STREAM GRADIENT & TOPOGRAPHIC PROFILE

Student Name: _____   Lab. Sec. # _____

1. Stream gradient increases where contour
   lines are close together when crossing
   a stream.   TRUE   FALSE

2. It is not necessary to know the scale
   of a map to determine stream gradient.   TRUE   FALSE

3. As the relief of an area decreases it
   is necessary to increase V.E. for a
   profile.   TRUE   FALSE

4. If a topographic map has a contour
   interval of 20', a highest contour
   line of 1120' and a lowest contour
   line value of 320', total relief
   on the map could exceed 800'.   TRUE   FALSE

5. When necessary to change V.E. on a
   topographic profile, the horizontal
   scale should be changed.   TRUE   FALSE

6. Streams on topographic maps do not
   always have contour lines crossing
   them.   TRUE   FALSE

7. - 10. Draw a profile for the map below:

1:24,000   C.I. 50'

900'
850'
800'
750'
700'
650'
600'
550'
500'
450'

1:24,000   C.I. 50'   V.E. = _____

179

EXERCISE 7-D

STREAM GRADIENT & TOPOGRAPHIC PROFILE

Student Name: _____ Lab. Sec. # _____

1. Contour line shape alone cannot be
   used to determine the direction of
   stream flow.   TRUE   FALSE

2. A stream gradient of 5.8'/mile means
   that the stream drops at a uniform rate.   TRUE   FALSE

3. Vertical exaggeration in profiles should
   be controlled by changing the value of
   the parallel lines.   TRUE   FALSE

4. Contour lines cross a stream at 1320' and
   1200' over a distance of 120 miles. What
   is the stream gradient?   _____

5. In areas where relief is 1000's of feet,
   a vertical exaggeration of nearer 10 x
   than 100 x is more desirable.   TRUE   FALSE

6. The horizontal scale on topographic maps
   cannot be changed to alter the V.E.   TRUE   FALSE

7. - 10. Draw a profile for the map below:

_____ 260'

_____ 240'

_____ 220'

_____ 200'

_____ 180'

_____ 160'

1:48,000   C.I. 20'   V.E. = _____

181

EXERCISE 7-E
STREAM GRADIENT & TOPOGRAPHIC PROFILE

Student Name: _____ Lab. Sec. # _____

1. The amount of stream gradient determines
   the speed of water flow.   TRUE   FALSE

2. If a stream has a gradient of 10'/mile,
   it could have a 6' water fall.   TRUE   FALSE

3. A topographic profile would be a straight
   line if no contours cross the profile line.   TRUE   FALSE

4. Contour lines cross a stream at 700'
   and 540' over a distance of 20 miles.
   What is the stream gradient?   _____

5. Hachured contour lines are not used
   when drawing topographic profiles.   TRUE   FALSE

6. The vertical scale of a topographic
   profile is the number of feet of
   change in _____ on the profile lines.

7. - 10. Draw a profile for the map below:

A ——— A'

                                                              640'
                                                              600'
                                                              560'
                                                              520'
                                                              480'
                                                              440'
                                                              400'

1:62,500   C.I. 40'   V.E. = _____

EXERCISE 7-F
STREAM GRADIENT & TOPOGRAPHIC PROFILE

Student Name: _____   Lab. Sec. # ____

1. The construction of dams & reservoirs
    has an effect on stream gradients.   TRUE   FALSE

2. It is possible for a stream with a
    10'/mi. gradient to drop less than
    1' over 3/4 mile.   TRUE   FALSE

3. The larger the V.E. for a topographic
    profile, the less it resembles the
    area it represents.   TRUE   FALSE

4. Contour lines cross a stream at 150'
    and 175' over a distance of 5 miles.
    What is the stream gradient?   _____

5. If no contour lines cross a stream on
    a topographic map, there is no gradient.   TRUE   FALSE

6. What does a V.E. of 10x actually mean?

    _____

    _____

7. - 10.   Draw a profile for the map below:

A —— A'

850   850   900

_____1100'

_____1050'

_____1000'

_____950'

_____900'

_____850'

_____800'

1:20,000   C.I. 80'   V.E. = _____

185

EARTH SCIENCE   LAB. EXERCISE #8

"INTRODUCTION TO MINERALS"

- -

OBJECTIVES:

Be Able to:

1. EXPLAIN THE SIGNIFICANCE OF MINERALS TO EVERYONE.

2. LIST THE 8 MOST COMMON ELEMENTS IN THE EARTH'S
    CRUST.

3. DISTINGUISH BETWEEN MINERALS AND ROCKS.

4. EXPLAIN THE RELATIONSHIP BETWEEN ELEMENTS,
    MINERALS, AND ROCKS.

5. DISCUSS THE APPROXIMATE MEMBER OF MINERALS PLUS
    THEIR METHODS OF FORMATION.

6. INDICATE SOME METHODS BY WHICH MINERAL IDENTIF-
    ICATION CAN OCCUR.

7. RECOGNIZE THE SEVEN UNUSUAL MINERAL PROPERTIES
    DISCUSSED IN THIS EXERCISE.

8. DEFINE THE GLOSSARY TERMS.

GLOSSARY TERMS

Mineral:

Fund Resources:

Flow Resources:

Rock:

Ore:

Inorganic:

Homogeneous:

Native Elements:

Silicates:

Nonsilicates:

Crystallize:

Hard Water:

Precipitates:

Diagnostic Traits:

Feel:

Diaphanity:

Taste:

Evaporites:

Smell:

Magnetism:

Tenacity:

HCl Reaction:

Carbonates:

HCl:

EXERCISE #8

INTRODUCTION TO MINERALS

Who care about minerals anyhow? We all should share an equal interest in the answer to this question! Minerals are the basis for our existence, and are the basic ingredients in everything from our foods to nearly all the materials surrounding you in this room. After a series of international crises involving the availability of natural resources in the 1970's and 1980's, perhaps mankind is finally becoming aware of the importance of natural resources in general, and minerals in particular.

Mineral resources exist in fixed or finite amounts in nature; consequently, they are referred to as "fund" or "non-renewable resources". Consequently, once they have been consumed they are gone forever! This is the cause for much widespread international concern, and has given rise to such policies as mandatory recycling and resource allocation restrictions. The location, thus availability, of minerals and natural resources has been the cause of nation's going to war! Are you aware of any such situations?

Solving mineral resource shortages, both present and future, will not be an easy task. Some of the possible solutions to an ever growing natural resource shortage will include:

    a.) Increased exploration for new deposits;
    b.) Better utilization of existing reserves;
    c.) Development of synthetic replacements.

Become familiar with the differences in the following terms:

NATURAL RESOURCE: "products of nature of value to man and available for his use".

FUND RESOURCES: "non-renewable resources such as ores and fossil fuels".

FLOW RESOURCES: " renewable resources such as air, water, forests, and crops".

MINERAL: " a solid, naturally occurring, homogeneous, inorganic, chemical element or compound".

ROCK: " an aggregate of one or more minerals".

MINERALS

An understanding of minerals is considered the single most important aspect of geology - the study of the physical earth. While minerals have such obvious economic importance as sources of metals for industry (ores), basic components of soils (sands and clays), and are the principal ingredients of many everyday items (glass, ceiling tile, and building materials), perhaps the best reason to understand them is to comprehend their role as basic building blocks of the solid rock of plant earth.

In simple terms, elements form minerals and minerals form rocks! Compare a mineral specimen with a rock specimen - what differences in composition can be seen?

The author of your textbook defines a mineral in a slightly different way: "any naturally occurring inorganic solid that is characterized by a definite chemical composition and by a specific and regular arrangement of atoms."

> How does a rock differ from a mineral?
> Could ice be considered a mineral?
> Why isn't coal a mineral?
> How about pearls?

Some minerals only contain 1 element and are called "native elements", can you give examples?

Relative Abundance of Earth's Crustal Elements	
Element	% of crustal Weight
Oxygen (O)	46.6
Silicon (Si)	27.7
Aluminum (Al)	8.1
Iron (Fe)	5.0
Calcium (Ca)	3.6
Sodium (Na)	2.8
Potassium (K)	2.6
Magnesium (Mg)	2.1
Remaining Elements	1.5
TOTAL	100%

(know)

Figure 38.

190

MINERAL OCCURRENCE

There are over 2000 separate minerals known to exist with new ones discovered nearly every year. The great majority of minerals, however, are very rare and less than 25 are considered common in nature. These 25 commonly occurring minerals are known as rock-forming minerals since they make up the bulk of earth's rocks. Furthermore, <u>Figure 38.</u> indicates the relatively simple <u>or</u> concentrated chemical composition of the earth's crustal minerals and rocks.

For <u>future</u> <u>reference</u> <u>it</u> <u>is</u> <u>important</u> <u>to</u> <u>remember</u> <u>the</u> <u>following</u>:

1) Oxygen and Silicon are the two most common elements in the earth's crust;

2) Minerals that contain Si and O in their chemical formula are referred to as <u>"Silicates"</u> -- this is the largest and most important group of earth minerals.

3) Feldspar ($KAlSi_3O_8$) and Quartz (SiO_2) are respectively the #1 and #2 most common minerals on earth;

4) All additional mineral groups are collectively referred to as the "nonsilicates". Although they contain important minerals, they are scarce by comparison to the silicates.

Minerals most commonly form in two very different and distinct environments:

a) when molten rock (lava or magma) cools thus allowing the elements to "crystallize";

and

b) when "hard water" on the earth's surface evaporates and dissolved elements are precipitated from the solution. <u>Which</u> <u>of</u> <u>these</u> <u>processes</u> <u>do</u> <u>you</u> <u>think</u> <u>has</u> <u>produced</u> <u>the</u> <u>majority</u> <u>of</u> <u>earth's</u> <u>minerals?</u>

<u>Why?</u>

191

IDENTIFICATION OF MINERALS

The absolute best method of mineral identification is through a sophisticated chemical analysis; however, such laboratory study is both slow and expensive. Fortunately the majority of common minerals have a few visible properties _or_ traits that are so unique that they are "diagnostic" for identification or recognition of their respective mineral. Consider the following 7 mineral properties that can be very helpful (perhaps even diagnostic) for mineral identification.

UNUSUAL PHYSICAL PROPERTIES OF MINERALS

1) FEEL: By merely feeling the surface of some minerals you may be confident of that minerals identity. Typical adjectives might be "abrasive", "gritty", "cold", "smooth", and so forth. How would you describe the feel of mineral specimen #16 in the kit provided?

2) OPTICS: The appearance of a mineral when light attempts to pass through it is referred to as diaphanity".

There are 3 types of diaphanity:

a) Transparent ... you can see through the mineral.

b) Translucent ....not transparent, but light passes through.

c) Opaque ....completely blocks light.

Find examples of all 3 types of diaphanity in the kit provided. Which diaphanity do you think is most common? What determines mineral diaphanity?

3) TASTE: Some minerals have very distinctive taste; examples include "salty", "sweet", "bitter", "alum", and so forth. A group of minerals known as "evaporties", because of their method of formation, are most typical of distinct taste. Because of sanitary considerations you may not desire to try this test! If this does not concern you, clean a corner of mineral specimen #10 and touch it to the tip of your tongue! (Please, no licking or "slobbering")! Result?

4. __SMELL:__   A few minerals emit a distinctive odor such as "rotten eggs", "musty", or "petroleum". A classic example is the smell of mineral #12 when it is damp (exhale deeply on it for several seconds then quickly smell the surface)! When people speak of the fresh smell outside after a summer rain, what do you suppose they are smelling?

5. __MAGNETISM:__ Some iron-rich minerals are magnetic enough to be detected with small magnets or iron filings. Use the small magnet provided to see this property in mineral specimen #3.

6. __TENACITY:__   When pressure is applied to a mineral it will respond in one of 3 ways: the mineral will thus be:

   a) __Brittle__ .... if it breaks;
   b) __Flexible__ ... if it bends & remains bent;
   c) __Elastic__ .... if it bends but returns to form.

   Which of these tenacity types do you think is most common?
   What are some variables of mineral tenacity?

7. __HCl ACID__
   __REACTION:__   A group of nonsilicate minerals known as "carbonates" react to hydrochloric acid -- HCl. When a small drop of this acid is placed on the surface of a carbonate mineral, a reaction occurs that can be either seen or heard. BE CAREFUL: The HCl acid is dangerous to eyes and clothes! Use only a small drop and wipe off with kleenex when finished. Use the HCl in the dropper bottle provided to see this reaction with mineral specimens #7 and #8. Are the reactions the same?

Additional mineral properties will be examined in the next laboratory exercise. If time permits, you may wish to examine additional examples of the above mineral traits provided in lab.

STUDY REVIEW: (Out-of-class)

1. Define mineral: _____
   _____
   _____

2. What is a diagnostic mineral property?
   _____
   _____
   _____

3. What are the silicates? _____
   _____

   Why are they so important? _____
   _____

4. What are the 8 most common elements in the earth's crust
   in order of abundance?

   _____ _____ _____ _____ _____

   _____ _____ _____

5. How do minerals form? _____
   _____
   _____

6. List a number of reasons why everyone has interest in
   minerals: _____
   _____
   _____--
   _____

7. Of what significance is feldspar and quartz?
   _____
   _____

8. What is the difference in rocks and minerals?
   _____
   _____

9. Why isn't the best method of mineral identification
   routinely used? _____
   _____

10. There are over _____ minerals of which less than
    _____ are common.

EXERCISE 8-A
INTRODUCTION TO MINERALS

Student Name: _____ Lab. Sec. # _____

1. It is more correct to refer to oil as
   a natural resource than a mineral
   resource.   TRUE   FALSE

2. Both $KAlSi_3O_8$ and SiO_2 are "silicates".   TRUE   FALSE

3. What is a basic difference in a rock

   and a mineral? _____

   _____

4. The most common element in the earth's continental

   crust is _____.

5. The most common mineral in the earth's continental

   crust is _____.

6. Explain the meaning of a diagnostic "mineral trait? _____

   _____

   _____

7. - 10. Using the mineral specimens provided, identify one
          of the 7 unusual physical properties covered in
          this lab. with each. (Use 4 different properties.)

7. Specimen displays: _____

8. Specimen displays: _____

9. Specimen displays: _____

10. Specimen displays: _____

Record your unknown set # _____

EXERCISE 8-B
INTRODUCTION TO MINERALS

Student Name: _____ Lab. Sec. # _____

1. All mineral resources are fund resources.   TRUE   FALSE

2. The "carbonates" are the largest and most
   important group of earth minerals.   TRUE   FALSE

3. What is a basic difference in a rock and a mineral?

   _____

   _____

4. The second most common element in the earth's continental

   crust is _____.

5. In terms of abundance, the mineral quartz is # _____ in

   the earth's continental crust.

6. Explain the meaning of a "diagnostic"
   mineral trait.

   _____

   _____

   _____

7. - 10.   Using the mineral specimens provided, identify one
           of the 7 unusual physical properties covered in
           this lab. with each. (Use 4 different properties.)

7. Specimen displays: _____

8. Specimen displays: _____

9. Specimen displays: _____

10. Specimen displays: _____

Record your unknown set # _____.

EXERCISE 8-C
INTRODUCTION TO MINERALS

Student Name: _____ Lab. Sec. # _____

1. Some minerals are flow resources.   TRUE   FALSE

2. There is more Calcium in the earth's
    continental crust than Iron.   TRUE   FALSE

3. What is a basic difference in a rock and
    a mineral? _____

    _____

4. What is a native element? _____

    _____

5. Name 2 methods by which minerals form:

    _____

    and _____.

6. Explain the meaning of a "diagnostic" mineral trait.

    _____

    _____

    _____

7. - 10.   Using the mineral specimens provided, identify one
            of the 7 unusual physical properties covered in
            this lab. with each. (Use 4 different properties.)

7. Specimen displays: _____

8. Specimen displays: _____

9. Specimen displays: _____

10. Specimen displays: _____

Record your unknown set # _____.

EXERCISE 8-D
INTRODUCTION TO MINERALS

Student Name: _____ Lab. Sec. # _____

1. It is possible for ice to be a mineral.   TRUE   FALSE

2. Two elements make up over 70% of weight
   of the earth's continental crust.   TRUE   FALSE

3. What is a basic difference in a rock and
   a mineral? _____

   _____

4. There are over _____ separate minerals known to exist but

   less than _____ are common.

5. What exactly is a "silicate" mineral? _____

   _____

6. Explain the meaning of a "diagnostic" mineral trait.

   _____

   _____

   _____

7. - 10.   Using the mineral specimens provided, identify one
           of the 7 unusual physical properties covered in
           this lab. with each. (Use 4 different properties.)

7. Specimen displays: _____

8. Specimen displays: _____

9. Specimen displays: _____

10. Specimen displays: _____

Record your unknown set # _____.

EXERCISE 8-E
INTRODUCTION TO MINERALS

Student Name: _____ Lab. Sec. # _____

1. All flow resources are renewable.   TRUE   FALSE

2. There is more oxygen in the earth's crust
    than <u>all</u> other elements combined.   TRUE   FALSE

3. Nonsilicate minerals are important but
    far less abundant than silicates.   TRUE   FALSE

4. What is a basic difference in a rock and a mineral?

    _____

    _____

    _____

5. An example of a native element is the

    mineral _____.

6. Explain the meaning of a "diagnostic" mineral trait.

    _____

    _____

    _____

7. - 10. Using the mineral specimens provided, identify one
            of the 7 unusual physical properties covered in
            this lab. with each. <u>(Use 4 different properties.)</u>

7. Specimen displays: _____

8. Specimen displays: _____

9. Specimen displays: _____

10. Specimen displays: _____

<u>Record your unknown set #</u> _____.

EXERCISE 8-F
INTRODUCTION TO MINERALS

Student Name: _____ Lab. Sec. # _____

1. It is possible to deplete flow resources
    but with proper management, it can be
    avoided.   TRUE   FALSE

2. One should expect to find more minerals
    which contain copper than magnesium.   TRUE   FALSE

3. What is a basic difference in a rock and a mineral?

    _____

    _____

4. Only <_____ of the earth's _____ known minerals
    are considered common in occurrence.

5. For minerals to form from "hard water", the water must

    _____.

6. Explain the meaning of a "diagnostic" mineral trait.

    _____

    _____

    _____

7. - 10. Using the mineral specimens provided, identify one
          of the 7 unusual physical properties covered in
          this lab. with each. (Use 4 different properties.)

7. Specimen displays: _____

8. Specimen displays: _____

9. Specimen displays: _____

10. Specimen displays: _____

Record your unknown set # _____.

EARTH SCIENCE   LAB. EXERCISE #9
"MINERAL PROPERTIES"

- -

OBJECTIVES:

Be Able To:

1. LIST THE 8 MAJOR PHYSICAL PROPERTIES OF
    MINERALS COVERED BY THIS EXERCISE.

2. EXPLAIN CRYSTAL FORM AND IDENTIFY THE 4
    TYPES DISCUSSED.

3. DISCUSS MINERAL COLOR AND EXPLAIN WHY IT
    IS MOST OFTEN AN UNRELIABLE PHYSICAL PROPERTY.

4. EXPLAIN AND DETERMINE MINERAL STREAK.

5. DESCRIBE AND RECOGNIZE THE 6 TYPES OF
    MINERAL LUSTER PRESENTED.

6. WRITE THE 10 MINERALS OF MOH'S HARDNESS
    SCALE IN ORDER, & EXPLAIN THE DETERMINATION
    OF HARDNESS.

7. DISTINGUISH BETWEEN MINERAL CLEAVAGE AND
    FRACTURE, AND RECOGNIZE THE VARIOUS FORMS
    DISCUSSED FOR EACH.

8. EXPLAIN THE MEANING OF SPECIFIC GRAVITY, AND
    THE USE OF "HEFTING" IN MINERAL IDENTIFICATION.

9. DEFINE THE GLOSSARY TERMS.

B.C. by permission of Johnny Hart and Field Enterprises, Inc.

GLOSSARY TERMS:

Cubic Crystal Form:

Octahedral Crystal Form:

Hexagonal Crystal Form:

Dodecahedral Crystal Form:

Color:

Streak:

True Color:

Streak Plate:

Luster:

Metallic:

Vitreous:

Silky:

Waxy:

Resinous:

Earthy:

Moh's Hardness:

Cleavage:

Basal:

Octahedral:

Rhombic:

Prismatic:

Fracture:

Conchoidal:

Hackly:

Fibrous:

Even:

Specific Gravity:

EXERCISE #9

MINERAL PROPERTIES

In this laboratory exercise you will examine the <u>major physical properties</u> of minerals used in their identification. It will become necessary first for you to learn the meaning of each category; second, you need to be able to distinguish between varieties within each mineral property category; and finally, you will need to recognize the physical mineral properties as shown by laboratory mineral specimens.

The physical properties of minerals to be examined in this laboratory exercise include:

1)	CRYSTAL FORM;	5)	HARDNESS;
2)	COLOR;	6)	CLEAVAGE;
3)	STREAK;	7)	FRACTURE;
4)	LUSTER;	8)	SPECIFIC GRAVITY;

- -

1.) <u>CRYSTAL FORM:</u> The crystal form of a mineral refers to the external form of the interior geometry. The crystal form is determined by the packing arrangement of the atoms in the compound. <u>Figure 39</u> demonstrates this relationship. Four common crystal forms of minerals are:

a) Cubic .... 3 planes, 90° angles;
b) Octahedral....... 4 planes; pyramid-like;
c) Hexagonal........ 6 sided; 60° angles;
d) Dodecahedral..... 6 planes; 12 sided.

<u>Are crystals common? Why?</u>

<u>If a mineral has cubic crystal form, why might it have a rectangular shape?</u>

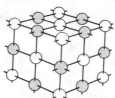

<u>Figure 39.</u>

2) **COLOR:** The meaning of color is obvious and results from the chemical composition _and_ crystal form of the mineral. Although minerals often display beautiful colors, a variety of color is possible within the same mineral species because of impurities and/or differences in structure. Where constant, mineral color can be diagnostic; too often, however, color is not a good diagnostic property!

What does the color red usually suggest about the composition of a mineral?

How about green?

Do you think color in native elements is more reliable than color for compounds?

Why?

3) **STREAK:** Streak is merely the color of a mineral in a powdered form; as such, this is the true color of a mineral! To obtain the streak of a mineral, it is rubbed across an unglazed piece of porcelain (streak plate). Any powdered residue is the streak of the mineral. DO NOT HOLD THE STREAK PLATE WHEN TESTING FOR STREAK! Place the streak plate on a hard surface in case the plate breaks!

What if minerals are harder than the streak plate?

Why are white and black streaks not considered diagnostic?

Look at mineral #2 in your kit -- its outward

color is _____ but its "true color" is

_____.

Why are several short streak tests from various areas of the mineral more reliable than 1 long streak test from a single area?

4) **Luster:** Luster refers to the quantity and quality of light <u>reflected</u> from the surface of a mineral; it should not be confused with mineral color! Specific types of mineral luster include:

    a) Metallic ..... clean, untarnished metal;
    b) Vitreous...... clean glass;
    c) Silky......... sheen of parallel fibers;
    d) Waxy.......... subdued glow (candles);
    e) Resinous...... amber-like colors;
    f) Earthy........ dull surface.

<u>What</u> <u>are</u> <u>some</u> <u>factors</u> <u>that</u> <u>influence</u> <u>the</u> <u>luster</u> <u>of</u> <u>a</u> <u>mineral?</u>

<u>Are</u> <u>weathered,</u> <u>or</u> <u>fresh</u> <u>mineral</u> <u>surfaces</u> <u>more</u> <u>reliable</u> <u>for</u> <u>luster</u> <u>description?</u> <u>Why?</u>

5) **HARDNESS:** Hardness is defined as the resistance to being scratched and is determined by the chemical bonding strength of the minerals atoms. Great variations in mineral hardness exist; this can therefore be a most diagnostic physical property! The German mineralogist, Friedrich Mohs, established a scale of increasing mineral hardness for unknown specimens to be compared with -- this is known as the "Moh's Hardness Scale". Any mineral on the scale will scratch a mineral with a lower number. <u>Note:</u> There is no numeric relationship to members of the scale! Diamond (#10) is not 10x harder than Talc (#1) -- it is 1000's of times harder! Learn the Moh's Hardness Scale as shown in <u>Figure</u> <u>40.</u>

[handwritten note: TAll GiRls can Flint And oTHer queer thing can do]

[handwritten note: Know]

"MOH'S HARDNESS SCALE"	
#1. Talc	#6. Orthosclase Feldspar
#2. Gypsum	#7. Quartz
#3. Calcite	#8. Topaz
#4. Fluorite	#9. Corundum
#5. Apatite	#10. Diamond

<u>Figure</u> <u>40.</u>

It is also helpful to remember the hardness of some common items such as: fingernail = 2.5; copper penny = 3.0; steel knife = 5.5; window glass = 6.0.

<u>Why</u> <u>is</u> <u>testing</u> <u>for</u> <u>the</u> <u>validity</u> <u>of</u> <u>diamond</u> <u>by</u> <u>scratching</u> <u>glass</u> <u>not</u> <u>a</u> <u>good</u> <u>test?</u>

<u>Why</u> <u>should</u> <u>a</u> <u>hardness</u> <u>test</u> <u>always</u> <u>be</u> <u>reversed?</u>

<u>What</u> <u>is</u> <u>the</u> <u>difference</u> <u>between</u> <u>scratching</u> <u>&</u> <u>crushing?</u>

6) **CLEAVAGE:** Cleavage is the tendency of minerals to break or "split" along definite planes; the shapes shown by cleavage may _or_ may not be the same as the mineral's crystal form! The cleavage or breaking pattern of a mineral is determined by the arrangement of atoms within the crystal form; planes of weakness occur between atoms—not through them! Learn the following types of mineral cleavage:

a) Basal...... flat, parallel sheets, 1 plane;
b) Cubic...... 90° angles, step-like; 3 planes;
c) Octahedral. pyramid-like; 4 planes;
d) Rhombic.... slanted cube; 3 planes;
d) Prismatic.. rhombic-like but only 2 planes.

What is an easy way to identify the presence of a cleavage plane?

How can a cleavage form be distinguished from a crystal form?

If unsure, which is more likely present? Why?

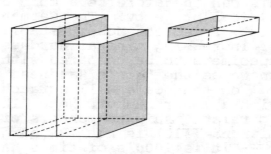
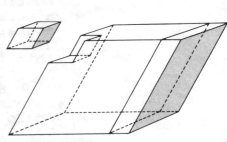

7) **FRACTURE:** Fracture refers to the appearance of a break that is not a cleavage plane; consequently, smooth (light reflecting) surfaces will not be present. Fracture is the opposite of cleavage -- a mineral can display both properties! Examples of mineral fracture include:

a) Conchoidal.... curved, smooth ripples;
b) Hackly........ rough to sharp surface;
c) Fibrous....... irregular fiber-like strands;
d) Even.......... smooth, undulating surface;

How can a mineral display both cleavage and fracture? What causes mineral fracture?

What is one of the oldest uses of fracture by mankind?

Why?

8) **SPECIFIC GRAVITY:** The specific gravity of any mineral is the ratio of its weight to the weight on an equal volume of water. A cubic foot of water weighs 62.4 lbs.; a cubic foot of Aluminum weighs 168.5 lbs., or 2.7 x as much; therefore, aluminum has a S.G. of 2.7x. Mineral specimens are most commonly "hefted" (tossed lightly in the hand) and then judged as to whether they feel heavier or lighter than expected for their size.

What factors influence the S.G. of a mineral?

What type of minerals will most often have the highest specific gravity?

To compare the S.G. of two different minerals both the _____ and _____ of each mineral must be considered.

Remember that it will not be necessary to learn all physical properties shown by each mineral. It is only necessary to learn the 1 or 2 outstanding, thus diagnostic traits! You do need to know all the categories and varieties, however, since different traits are important for different minerals.

Examine the physical properties of minerals displayed by the lab. specimens provided. Your laboratory instructor will indicate the time allotment for your study.

SELF-STUDY (Using a mineral kit in lab, the library, or study center, associate each of the following with 1 or 2 diagnostic properties.

MINERAL*	DIAGNOSTIC PROPERTIES

Handwritten annotation above Mineral column: "color"

1. Galena..... Gray .... Cold metal feel | Cubic *(cleavage)* | LEAD *(gravity)*
2. Sphalerite.............. SulFer | Streak yellow | Luster Resinous | gravity ZINC
3. Magnetite................ magnetic *Iron*
4. Hematite.... *scabe* Red .... Streak DARK Red | gravity Iron
5. Bauxite................. Lows specific Gravity
6. Pyrite.. (Fools gold)... Cleavage Cubic | gravity Iron *(no gold)*
7. Calcite................. Hardness #7 | Cleavage Rhombic
8. Dolomite................ Slow HCl Test.
9. Fluorite................ Octahedral | Hardness #4
10. Halite................. SAlt Cubic
11. Selenite Gypsum........ Hardness #2 | Cleavage BASAl | *Flexable*
12. Kaolinite............. Sticks to tongue (Chalk) musty odor.
13. Olivine............... olive Green | grav. Iron
14. Hornblende............ Black | Luster Silky | Cleavage Prismatic
15. Mica.................. Cleavage BASAl | BREAKS off EASY
16. Talc................. Green Soapy Feel | Hardness #1 | Fracture even
17. Feldspar............. Hardness #6 | Cleavage Prismatic
18. Quartz............... hexagonal | Luster vitreous | Hard #7 | Fracture conchodial
19. Garnet............... sharp Abrasive feel | Fracture HACKly | Gravity Iron
20. Chrysotile Serpentine... Fracture Fibrous | Asbestis

NOTE: Numbers correspond to mineral numbers in study kit.

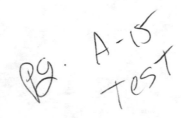

Handwritten in left margin: A150 242

216

EXERCISE 9-A
MINERAL PROPERTIES

Student Name: _____ Lab. Sec. # _____

1. The majority of minerals you are likely
   to find will display well formed crystals. (TRUE) FALSE

2. The "true color" of a mineral is its: _streak_ .

3. According to the Moh's Hardness Scale
   quartz is twice as hard as calcite.   TRUE (FALSE)

4. A mineral specimen can display both
   cleavage and fracture.   TRUE   FALSE

5. Specific Gravity and luster are more
   reliable diagnostic physical properties
   than feel and diaphanity.   TRUE   FALSE

6. It is usually necessary to learn 5-10
   physical properties of a mineral to
   make its identification.   TRUE (FALSE)

Using the set of unknowns you have been given:

7. What is the hardness of specimen A?   _____

8. What is the crystal form of specimen B?   _____

9. Identify the cleavage or fracture of
   specimen C: _____

10. Identify the luster of specimen D: _____

Record your unknown set # _____.

EXERCISE 9-B
MINERAL PROPERTIES

Student Name: _____   Lab. Sec. # _____

1. A mineral crystal with more than 5
    sides must be dodecahedral.   TRUE (FALSE)

2. Mineral color and streak may be the
    same, or they may be different.   (TRUE) FALSE

3. Moh's Hardness Scale does not suggest
    fluorite is twice as hard as gypsum.   TRUE (FALSE)

4. It is not possible for a mineral to
    display both cleavage and fracture.   (TRUE) FALSE

5. Composition alone determines the specific
    gravity of minerals; minerals that contain
    metals will always be heavier than those
    which do not.   TRUE FALSE

6. While most often unreliable, in some
    cases, mineral color can be a useful
    physical property.   (TRUE) FALSE

- -

Using the set of unknowns you have been given:

7. What is the hardness of specimen A?   _____

8. What is the crystal form of specimen B?   _____

9. Identify the cleavage or fracture of
    specimen C: _____

10. Identify the luster of specimen D: _____

Record your unknown set # _____.

221

EXERCISE 9-C
MINERAL PROPERTIES

Student Name: _____   Lab. Sec. # _____

1. If a mineral displays a crystal form, its cleavage must be that same form.   TRUE   FALSE

2. Mineral color is a more reliable physical property than streak.   TRUE   (FALSE)

3. The mineral quartz will scratch glass.   (TRUE)   FALSE

4. If a mineral displays fracture, smooth planes that reflect light will be present.   (TRUE)   FALSE

5. _hefted_ refers to a simple test to determine the approximate specific gravity of a mineral.

6. Color, specific gravity, and cleavage are all influenced by mineral composition and internal atomic structure.   (TRUE)   FALSE

- -

Using the set of unknowns you have been given:

7. What is the hardness of specimen A?   _____

8. What is the crystal form of specimen B?   _____

9. Identify the cleavage or fracture of specimen C: _____

10. Identify the luster of specimen D: _____

Record your unknown set # _____.

EXERCISE 9-D
MINERAL PROPERTIES

Student Name: _____ Lab. Sec. # _____

1. A crystal is growing out of a rock.
    Only part of the total crystal can
    be seen, but there are at least
    5 planes. The crystal form is most   _____
    likely:

2. It is not possible for a given mineral
    species (i.e. quartz) to display a
    variety of colors.   (TRUE) FALSE

3. The softest known mineral is _TALC_____

    while the hardest is __Diamond__.

4. Mineral cleavage occurs as atoms them-
    selves are split; fracture occurs when
    atoms are separated from each other.   TRUE (FALSE)

5. The reference for determination of

    specific gravity is _____,

    which has a S.G. of 1.

6. It is more desirable to use a weathered
    surface than a fresh surface to
    determine luster.   TRUE (FALSE)

- -

Using the set of unknowns you have been given:

7. What is the hardness of specimen A?   _____

8. What is the crystal form of specimen B?   _____

9. Identify the cleavage or fracture of
    specimen C: _____

10. Identify the luster of specimen D: _____

Record your unknown set # _____.

225

EXERCISE 9-E
MINERAL PROPERTIES

Student Name: _____   Lab. Sec. # _____

1. It is possible to distinguish between
    octahedral and hexagonal crystals based
    on degree of angles alone.   TRUE   FALSE

2. Color alone can never be considered a
    diagnostic property.   (TRUE) FALSE

3. The average fingernail can scratch at
    least 2 minerals on the Moh's Hardness
    Scale.   (TRUE) FALSE

4. The basic difference in rhombic and
    prismatic cleavage is the number of
    planes.   (TRUE) FALSE

5. Indian arrowheads typically have a
    series of smooth, C-shaped depressions
    along their edges; these broken edges
    are actually _Conchoidal_ fractures.

6. The majority of minerals can be identified
    if only 2 or 3 of their most diagnostic
    physical properties are known.   (TRUE) FALSE

- -

Using the set of unknowns you have been given:

7. What is the hardness of specimen A?   _____

8. What is the crystal form of specimen B?   _____

9. Identify the cleavage or fracture of
    specimen C: _____

10. Identify the luster of specimen D: _____

Record your unknown set # _____.

EXERCISE 9-F
MINERAL PROPERTIES

Student Name: _____ Lab. Sec. # _____

1. Minerals are more likely to display cleavage or fracture than crystal form.　　　　　　TRUE　FALSE

2. The most common mineral streaks are white and black; these are thus not diagnostic.　　　　　TRUE FALSE

3. Basically, the crushing of a mineral is the same as its hardness.　　　　TRUE　FALSE

4. If a mineral displays cleavage, it will show that shape regardless of how many times the specimen is broken.　　TRUE　FALSE

5. Mineral luster is not related to mineral color for most specimens.　　TRUE　FALSE

6. The more physical properties you can determine for an unknown mineral, the more reliable the identification of the specimen.　　TRUE　FALSE

- -

Using the set of unknowns you have been given:

7. What is the hardness of specimen A?　　_____

8. What is the crystal form of specimen B?　　_____

9. Identify the cleavage or fracture of specimen C:　　_____

10. Identify the luster of specimen D:　　_____

Record your unknown set # _____.

EARTH SCIENCE   LAB. EXERCISE #10

"IDENTIFICATION AND USE OF MINERALS"

- -

OBJECTIVES:

Be Able to:

1. ASSOCIATE THE PHYSICAL PROPERTIES OF MINERALS
    FROM EXERCISES 8 AND 9, WITH THE 20 MINERALS
    BEING STUDIED, FOR IDENTIFICATION PURPOSES.

2. DEFINE AND IDENTIFY THE 5 MINERAL ORES BEING
    STUDIED.

3. IDENTIFY THE IMPORTANCE OF EACH OF THE 20
    MINERALS BEING STUDIED.

4. LIST THE MAJOR ROCK FORMING MINERALS BEING
    STUDIED.

5. CORRECTLY IDENTIFY THE 20 MINERAL SPECIMENS
    BEING STUDIED FROM VISUAL EXAMINATION OF AN
    UNKNOWN, AND/OR IF GIVEN THE MAJOR TRAITS.

6. DEFINE THE GLOSSARY TERMS.

B. C. by permission of Johnny Hart and Field Enterprises, Inc.

GLOSSARY TERMS:

Ore Minerals:

Rock Forming Minerals:

Lodestone:

Pisolitic Structure:

Fool's Gold:

Lime:

Double Refraction:

Iceland Spar:

Ag. Lime:

Fluxing Agent:

Emulsifier:

Satin Spar:

Filler:

Muscouite:

Biotite:

Isinglass:

#1 Mineral in Crust:

#2 Mineral in Crust:

Semi-Precious Gem:

Mica Book:

Herkimer Diamonds:

Carcinogenic:

EXERCISE #10

IDENTIFICATION AND USE OF MINERALS

This laboratory exercise is designed to utilize the information on mineral properties introduced in exercises #8 and #9 so that you will associate diagnostic physical properties with the correct mineral specimens for the purpose of identification. A listing of economic uses and other important facts for the 20 mineral specimens you are learning is included.

It is convenient to divide the 20 minerals we are covering in Earth Science into 3 groups as follows:

1. **ORE MINERALS:** Mined as a source for 1 or more metals; Kit specimens of the ores are: 1. Galena; 2. Sphalerite; 3. Magnetite; 4. Hematite; and 5. Bauxite.

2. **ROCK FORMING MINERALS:** Most abundant thus form most of earth's rock; Kit specimens of the rock forming minerals include: 7. Calcite; 8. Dolomite; 13. Olivine; 14. Hornblende; 15. Mica; 17. Feldspar; and 18. Quartz.

3. **SPECIAL INTEREST MINERALS:** Specimens with unusual properties and/or importance; Kit specimens in this group are: 6. Pyrite; 9. Fluorite; 10. Halite; 11. Selenite Gypsum; 12. Kaolinite; 16. Talc; 19. Garnet; and 20. Chrysotile Serpentine.

MINERAL GROUPS		
GROUP	COMPOSITION	EXAMPLE
Native Elements......Single Elements........		Au (Gold)
Oxides.... Oxygen + 1 or more Elements......		Fe_2O_3 (Hematite)
Sulfides.. Sulfur + 1 or more Elements......		FeS_2 (Pyrite)
Sulfates...SO_4 + 1 or more Elements.....		$CaSO_4$ $2H_2O$ (Gypsum)
Halide.....Cl or Fl + 1 or more Elements....		CaF_2 (Fluorite)
Carbonates..CO_3 + 1 or more Elements.......		$CaCO_3$ (Calcite)
Silicates..SiO_2 + 1 or more Elements..		$KAlSi_3O_8$ (K Feldspar)

Figure 41.

MINERAL DISCUSSIONS:

For each of the 20 minerals you are learning to identify, the following outline will be used to list the most significant information:

Mineral Number, Name, and Formula
Importance

Diagnostic Traits

#1 GALENA   PbS

Ore of Lead

Cubic Crystal Form, Cubic Cleavage, Metallic Luster, High Specific Gravity.

NOTES: _____

_____

_____

#2 SPHALERITE   ZnS

Ore of Zinc

Yellow Streak, Rotten Egg (Sulfur) Odor, Resinous Luster, Dodecahedral Cleavage.

NOTES: _____

_____

_____

#3 MAGNETITE Fe_3O_4

Ore of Iron (Scarce)

Metallic Luster, Magnetic, High S.G., "Lodestone", Dark Color.

NOTES: _____

_____

_____

#4 HEMATITE Fe_2O_3

Ore of Iron

Red Streak, Dull Luster, High S.G.

NOTES: _____

_____

_____

#5 BAUXITE

Ore of Aluminum

Low S.G., Dull Luster, Pisolitic Structure.

NOTES: _____

_____

_____

#6 PYRITE FeS_2

Referred to as "Fool's gold"; mined only in a few countries

Metallic Luster, Brassy-gold Color, Cubic Crystal Form and
Cleavage, Black Streak, Sulfur Odor on Streak Plate.

NOTES: _____

_____

_____

#7 CALCITE $CaCO_3$

Source of Calcium ("lime") for soil treatment; an ingredient
in Portland Cement, forms the rock limestone.

Hardness of 3, Rhombic Cleavage, Translucent to Transparent
Diaphanity ("double refraction") = Iceland Spar, Fast Reaction
to HCl (hydrochloric acid).

NOTES: _____

_____

_____

#8 DOLOMITE $CaMg(CO_3)_2$

Source of Calcium ("lime") for soil treatment = Ag. Lime.

Slow Reaction to HCl (hydrochloric acid), Pink and Purple Colors Common, Granular Look/Feel.

NOTES: _____

_____

_____

#9 FLUORITE CaF_2

Source of Fluorine for oral hygiene; fluxing agent; HF (hydrofluoric acid) for etching glass.

Hardness of 4, Octahedral Cleavage, Cubic Crystal Form, Vitreous Luster, Soft (Pastel) Colors Common = Green, Yellow, Blue, Lavender, etc...

NOTES: _____

_____

_____

#10 HALITE $NaCl$

Source of Chlorine for cleaning agents (i.e. chlorox bleach), also, simple table salt.

Cubic Crystal Form and Cleavage, Transparent to Translucent Diaphanity, "Salty Taste", Vitreous Luster.

NOTES: _____

_____

_____

#11   SELENITE GYPSUM $CaSO_4 \cdot 2H_2O$

Used to make dry wall (sheet rock or plaster board), ingredient in many plasters, "emulsifier" in some foods, livestock feed supplement.

Hardness of 2, Basal Cleavage, Transparent to Translucent Diaphanity, Flexible Tenacity, Vitreous Luster, Fibrous Fracture ("Satin Spar") on occasions.

NOTES: _____

_____

_____

#12   KAOLINITE $H_4Al_2Si_2O_9$

Used in manufacture of fine china and ceramics, paper, paints, and certain plastics.

Low Hardness, Dull Luster, White Color, Chalky Feel and Look, Musty Odor when Damp (breathe on it).

NOTES: _____

_____

_____

#13   OLIVINE $(Mg,Fe)_2SiO_4$

Rare mineral in earth's crust, common mineral in certain types of Meteorites, forms most commonly deep below earth's surface, no economic value.

Green Color, Granular Texture, Vitreous Luster, High S.G.

NOTES: _____

_____

_____

#14 HORNBLENDE   (SEE FORMULA BELOW)

Very common mineral in earth's crust, a principle component of such rocks as granite, no economic value.

Black to Dark Green Color, Silky Luster, Prismatic Cleavage.

Formula: $Ca_2Na(Mg,Fe_2)_4 (Al,Fe_3,Ti) (Al,Si)_8 O_{22} (O,OH)_2$

NOTES: _____

_____

_____

#15 Mica (Muscovite) $KAl_2(AlSi_3O_{10}) (OH)_2$

Two distinct varieties: Dark = Biotite & Light = Muscouite, used as an electrical insulator, filler in paints and tiles, "formica", once known as "isinglass" when used like glass, used in the movie industry as snow.

Basal Cleavage, Elastic Tenacity, Vitreous Luster, Low Hardness, Translucent to Transparent Diaphanity.

NOTES: _____

_____

_____

#16 TALC $Mg_3(Si_4O_{10}) (OH)_2$

Used as a filler, also in ceramics manufacturing, an ingredient in rubber, plastics and some lubricants, expensive non-reactive table tops (i.e. chemistry work areas), and powders such as talcum and baby powder.

Hardness of 1, Soapy or Greasy Feel, Commonly Greenish in Color.

NOTES: _____

_____

_____

#17   FELDSPAR (Orthoclase) $KAlSi_3O_8$

Most common mineral in the earth's continental crust, upon decomposition it yields the mineral kaolinite and other clay minerals which make up the clays of soils, also used to manufacture cheap glass products, and inexpensive ceramics.

Hardness of 6, Prismatic Cleavage, Typically an Off-White to Orangish Color.

NOTES: _____

_____

_____

18   QUARTZ SiO_2

Second (2nd) most common mineral in the earth's crust, used in electronics, occurs as the semi-precious gem stones (Amethyst, Rose, Citrine, Smoky, Opal, Agate, Sard, Chalcedony, Onyx, Jasper, Bloodstone, etc.), also used in glass manufacturing, and as an abrasive (sandpaper).

Hardness of 7, Hexagonal Crystal Form, Conchoidal Fracture, Transparent to Translucent Diaphanity, Vitreous Luster.

NOTES: _____

_____

_____

19   GARNET (Almandite) $Fe_3Al_2(SiO_4)_3$

Semi-precious gem stone, used as an abrasive (emery board).

Dodecahedral Crystal Form, Hackly Fracture, Most Common Varieties are Reddish in Color, High Hardness (over 8), Vitreous Luster.

NOTES: _____

_____

_____

#20 SERPENTINE (Chrysotile) $Mg_3Si_2O_5(OH)_4$

Most important type of asbestos, usage has been greatly reduced since it was found to be **"carcinogenic"**.

Fibrous Fracture, Greenish-white in Color, Waxy Luster.

NOTES: _____

_____

_____

_____

Be sure to examine a number of lab. specimens to see a variety of "Looks" the 20 mineral specimens can display!

MATCHING EXERCISE:

H	Galena	A.	Fool's Gold
K	Sphalerite	B.	Major Rock Former - Granite
N	Magnetite	C.	Iron Ore - common
C	Hematite	D.	Cheap Glass - Ceramics
S	Bauxite	E.	Limestone - Portland Cement
A	Pyrite	F.	Asbestos - Carcinogenic
Q	Calcite	G.	Common Salt - Bleaches
Q	Dolomite	H.	Lead Ore
P	Fluorite	I.	Electronics - Semi-Precious Gems
G	Halite	J.	Fine China/Ceramics
M	Selenite Gypsum	K.	Zinc Ore
J	Kaolinite	L.	Insulator - Filler
T	Olivine	M.	Dry Wall - Plasters
B	Hornblende	N.	Iron Ore - Lodestone
L	Mica	O.	Semi-Precious Gem - Abrasive
R	Talc	P.	Fluxing Agent - HF
B	Feldspar	Q.	Agricultural Lime
D	Quartz	R.	Powders - Lubricants
O	Garnet	S.	Aluminum Ore
F	Chrysotile Serpentine	T.	Meteorite Occurrence-Scarce

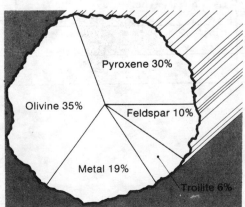

Olivine 35%
Pyroxene 30%
Feldspar 10%
Metal 19%
Troilite 6%

<----- **WHAT IS THIS?**

241

INDEPENDENT STUDY:

No do
Have to Know

PbS

ZnS

F₃64

Fe₂O₃

FeS₂

CaCO₃

Ca Mg O₃

NaCl

MINERAL NAME	IMPORTANCE	DIAGNOSTIC TRAITS	
1. Galena	Lead ore	metallic Luster Black streak CLEAVAGE 8 Cubic	
2. Sphalerite	Zinc ore Galvinizing	yellow streak Sulfur Smell S.G. Zinc	
3. Magnetite	High grade Iron ore	magnetic S.G. IRON	
4. Hematite	Iron ore	STREAK 8 DARKRED S.G. IRON	Red color
5. Bauxite	Aluminum ore	Psiolitic 8 Peanuts S.G. Low Aluminum ore	
6. Pyrite	Sulfuric Acid	Brassy color CLEAVAGE 8 cubic	Black Streak
7. Calcite	Soud Freatment	HARDNESS #3 CLEAVAGE 8 Rhombic	
8. Dolomite	Lime Agric.	Slow HCl Test	
9. Fluorite	Flux remove imperfects in steel	Hardness 8 #4 any color octahedral	
10. Halite	rock salt Sodium & Chlorine	salt, CUBIC	
11. Selenite Gypsum	DRYwall sheet rock	CLEAVAGE 8 BASAl HARDNESS 8 #2 flexable TeNasity	
12. Kaolinite	Porcelin China	Luster8earthy Fractureg even	sticks to tongue musty order
13. Olivine	Universal Rock Former	color 8 green S.G. IRON	
14. Hornblende	ROCK Former	color 8 Black Luster8 silky	Prismatic
15. Mica	electronic Insulation	Breaks off easy elastic Tenasity	basal cleavage
16. Talc	powder baby powder	Luster8waxy soapy feel Hardness8 #1	
17. Feldspar	Cheap glass & ceramics	Hardness # 6 CLEAVAGE 8 Prismatic	
18. Quartz	electronics	Luster8 vitreous HARDNESS 8 #7 Hexaginale	
19. Garnet	gem stone Abrasive	Fracture 8 hacky/Ruby red color S.G. IRON dorhexadedron	
20. Serpentine	Asbetas	FRACTURE = Fibrous (Aspestits)	

G-3

LiBRARY

STUDY SK-95 #31

EXERCISE 10-A
IDENTIFICATION & USE OF MINERALS

Student Name: _____ Lab. Sec. # _____

- -

Using the set of minerals you have been given, identify the
unknowns, name a major diagnostic property, and indicate its
economic use or importance.

- -

	MINERAL NAME	DIAGNOSTIC TRAIT	USE OR IMPORTANCE
A.	Galena	Cubic	Lead ore
B.	Sphalerite	Solfur	galvenizing
C.	Magnetite	Magnetic	High grade Iron ore
D.	Hematite	Red streak	Iron
E.	Bauxite	Low Aulminum ore	Aluminum ore
F.	Pyrite	Brassy color	Sulfuric Acid
G.	Calcite	Slow Hcl Test	Soil treatment
H.	Domolite	High Hcl Test	Lime Ag. use
I.	Florite	Cubic	Removes Imperitits from steel
J.	Halite	Salt taste	Rock salt

BONUS:

_____ _____ _____

<u>Record your unknown set #</u> _____

EXERCISE 10-B
IDENTIFICATION & USE OF MINERALS

Student Name: _____ Lab. Sec. # _____

- -

Using the set of minerals you have been given, identify the
unknowns, name a major diagnostic property, and indicate its
economic use or importance.

- -

	MINERAL NAME	DIAGNOSTIC TRAIT	USE OR IMPORTANCE
A.	Selenite Gypsum	Hardness #2	Drywall
B.	Kaolinite	musty odor	Porcelin / china
C.	Olivine	greenish color	Rock Former
D.	Hornblende	Silky Luster	Rock Former
E.	Mica	basal Fracture	electronic Insulation
F.	Talc	#1 hardness	Powders
G.	Feldspar	6# hardness	Cheap glass
H.	quartz	7# hardness	electronics
I.	Garnet	Ruby Red Color	Abrassive's
J.	Serpentine	Fiberous	Asbestas

BONUS:

_____ _____ _____

Record your unknown set # _____

EXERCISE 10-E
IDENTIFICATION & USE OF MINERALS

Student Name: _____ Lab. Sec. # _____

- -

Using the set of minerals you have been given, identify the unknowns, name a major diagnostic property, and indicate its economic use or importance.

- -

	MINERAL NAME	DIAGNOSTIC TRAIT	USE OR IMPORTANCE
A.	Galena	Cubic	Lead ore
B.	Sphalrite	Sulfar smell	Zinc ore
C.	~~Pyrox~~ MAGNAtite	MAGNetic	High GrAde Iron ore
D.	HemAtite	DARk Red STReak	~~Lead~~ Iron ore
E.	~~magnesa~~ Bauxite	LOw Aluminum ore	Aluminum ore
F.	Pyrite	Brassy color	Rock Forme
G.	~~Dolomite~~ calcite	Rhombic	Soil Treatment
H.	Dolomite	Slow Hcl Test	Ag. use
I.	Florite	#4 Hardness	flux removes impertics From steel
J.	HAlite	SAlt tafe	SAlt Rock

BONUS:

_____ _____ _____

Record your unknown set # _____

EXERCISE 10-F
IDENTIFICATION & USE OF MINERALS

Student Name: _____   Lab. Sec. # _____

- -

Using the set of minerals you have been given, identify the
unknowns, name a major diagnostic property, and indicate its
economic use or importance.

- -

	MINERAL NAME	DIAGNOSTIC TRAIT	USE OR IMPORTANCE
A.	Selenite Gypsum	Basal	Sheet Rock
B.	Kaolinite	Musty odor	china
C.	Olivine	Greenish	Rock Former
D.	Hornsblende	Silky	Rock Former
E.	Mica	basal	electronic Ins.
F.	Talc	#1 hardness	Powders
G.	Feldspar	Prismatic	cheap Glass
H.	Quartz	Hexagonal	electronics
I.	Garnet	Ruby red color	Abrasive
J.	Serpentine	Fiberous	Asbetas

BONUS:

_____   _____   _____

_____   _____   _____

Record your unknown set # _____

EARTH SCIENCE   LAB. #11
"INTRODUCTION TO ROCKS AND ROCK CYCLE"

- -

OBJECTIVES:

Be Able To:

1. EXPLAIN THE DIFFERENCES IN ROCKS AND MINERALS.

2. DEFINE THE TERM "ROCK".

3. DISCUSS THE RELATIONSHIP BETWEEN ROCKS
    AND MINERALS.

4. DEFINE AND EXPLAIN THE MEANING OF "MONO-
    MINERALIC ROCK", AND GIVE EXAMPLES.

5. DESCRIBE THE DIFFERENCE IN APPEARANCE OF
    ROCKS AND MINERALS; MONOMINERALIC AND
    OTHER ROCKS.

6. EXPLAIN HOW ROCKS ARE CLASSIFIED FOR
    IDENTIFICATION.

7. DIAGRAM AND EXPLAIN ALL ASPECTS OF THE
    "ROCK CYCLE".

8. DEFINE THE GLOSSARY TERMS.

B.C. by permission of Johnny Hart and Field Enterprises, Inc.

GLOSSARY TERMS:

Rock:

Aggregate:

Monomineralic:

Igneous:

Sedimentary:

Metamorphic:

Magma:

Lava:

Sediment:

Precipitates:

Metamorphism:

Lithification:

Detritus:

Weathering:

Erosion:

Deposition:

Crystallization:

Cementation:

Compaction:

Dessication:

Radiogenic Heat:

1st. Order Rocks:

2nd. Order Rocks:

3rd. Order Rocks:

EXERCISE #11
INTRODUCTION TO ROCKS AND ROCK CYCLE

It has already been stated that the principal components of the solid earth are minerals; if so, then the rocks are the form they take. Be sure to compare the following definition of a rock with that for a mineral covered earlier:

ROCK: An aggregate (mixture) of one or more minerals consolidated into a solid form.

A simple analogy may help understand the relationship between rocks and minerals: think of minerals as the individual vegetables in a tossed salad. The salad represents a separate item (the rock) which is composed of smaller, distinct individual vegetables (the minerals).

Another difference in rocks and minerals is the homogenity of composition. Minerals have exact compositions as indicated by the chemical formula for each mineral compound. Rocks, on the other hand, are identified largely by what minerals are present, but the amounts of the minerals can vary greatly from rock to rock.

A number of rocks are "monomineralic"; that is, they are composed of only 1 mineral with the balance of the rock being impurities. Examples of monomineralic rocks you will be studying are:

        #23   Syenite.........................Feldspar-rich
        #31   Sandstone (some)...............Quartz-rich
        #33   Limestone.....................Calcite-rich
        #39   Quartzite.....................Quartz-rich
        #40   Marble........................Calcite-rich

Examine several of the rocks in the kit provided and compare them with the minerals to see their visual compositional differences. Also compare the difference in appearance of the monomineralic rocks with any/all of the other rocks. Conclusion?

ROCK CYCLE

Initially it is not obvious as to how rocks can be classified or separated into groups. You might suspect that mineral composition alone would be an adequate criteria; however, rocks that have virtually identical mineral composition often display widely different structural appearances and had different origins. Consequently, a method of classifying rocks based on a combination of their a.) origin, b.) composition, and c.) individual structural features has been developed. Within this system all rocks are placed within 1 of 3 categories based on their origin (method of formation):

IGNEOUS: Rocks formed from the solidification of molten rock (magma <u>or</u> lava).

SEDIMENTARY: Rocks formed from the solidification of weathered debris (sediment), organic matter, or minerals precipitated from evaporating water.

METAMORPHIC: Rocks formed from the alteration of any type of previous existing rock, due to heat, pressure, or chemical change, without melting or breaking the rock.

Consider the 3 origins of rock as shown in <u>Figure 42.</u>

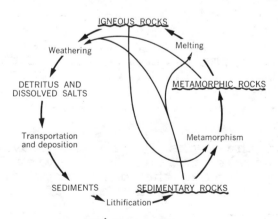

<u>Figure 42.</u>

FILM: "THE ROCK CYCLE"

Discussion Notes: _____

_____

_____

_____

_____

_____

_____

_____

_____

===

| Review of Rock Cycle |

Inside of earth

Breakdown particles at surface

movement of sediments

IGNEOUS ROCK

Weathering → Cooling/Crystallization

Erosion → Magma/Lava

Deposition → Melting[2]

Lithification[1] → METAMORPHIC ROCK

Settling down sediments

SEDIMENTARY ROCK → Heat/Pressure/Chemistry → Metamorphism

most common on surface

---

[1] Lithification by:

  a.) Cementation*
  b.) Compaction
  c.) Dessication ~Dissolving

[2] Melting by:

  a.) Radiogenic Heat
  b.) Friction

---

* Most common minerals responsible for cementation include: Quartz, Calcite, Dolomite, and Pyrite.

ROCK CYCLE EXERCISE

List the steps involved in a piece of the Igneous Rock granite becoming the Metamorphic Rock gneiss: _____

_____

_____

List the possible steps involved when mineral material within the Metamorphic Rock schist becomes part of the Sedimentary Rock sandstone.

_____

_____

_____

List all steps involved when a piece of the Sedimentary Rock limestone passes through all stages of the rock cycle:

_____

_____

_____

_____

_____

ROCK CYCLE QUESTIONS:

1. What type of rocks are most common on the earth's surface? _____ Why? _____
    _____

2. What type of rocks are least stable at the earth's surface? _____ Why? _____
    _____

3. Associate the following terms with the 3 categories of rocks:
    a. Primary (1st order)..... _____
    b. Secondary (2nd order)... _____
    c. Tertiary (3rd order).... _____

    Why are these appropriate? _____
    _____
    _____

4. What type of rock is most common for the earth as a total

    planet? _____

STUDENT EXERCISE: (OUT-OF-CLASS)

Use the space provided to construct your own version of the "rock cycle". Be sure to include <u>all</u> possible steps and show <u>all</u> possible paths of origin!

© Publishers Newspaper Syndicate 1966

B.C. by permission of Johnny Hart and Field Enterprises, Inc.

EXERCISE 11-A
INTRODUCTION TO ROCKS AND ROCK CYCLE

Student Name: _____ Lab. Sec. # _____

- -

1. It is more correct to say minerals
    form rocks than visca versa.   TRUE   FALSE

2. The variation between 2 pieces of
    the same mineral is likely to be
    greater than the variation between
    2 pieces of the same rock.   TRUE   FALSE

3. Rocks are classified based solely on
    their mineral composition.   TRUE   FALSE

4. It is possible for Sedimentary Rocks to
    form directly from Igneous Rocks.   TRUE   FALSE

5. Sedimentary Rocks are more stable at the
    earth's surface than Metamorphic Rocks.   TRUE   FALSE

6. Give an example of a monomineralic
    rock and its dominant mineral:

    _____ and _____.

7. What is the general difference in the appearance
    of most rocks compared to most minerals?

    _____

    _____

8. Heat and pressure is involved in the
    formation of ? rocks.   _____

9. The most common rock type <u>on</u> earth is: _____.

10. What are the 3rd. order or "derived rocks?   _____

EXERCISE 11-B
INTRODUCTION TO ROCKS AND ROCK CYCLE

Student Name: _____   Lab. Sec. # _____

- -

1. Minerals are more important to the
    formation of rocks than visca versa.   TRUE   FALSE

2. There is usually more difference in
    the appearance of 2 pieces of the same
    rock than 2 pieces of the same mineral.   TRUE   FALSE

3. Method or origin of formation is an
    important consideration when class-
    ifying rocks.   TRUE   FALSE

4. It is not possible for Metamorphic
    Rocks to form from both Igneous and
    Metamorphic Rock.   TRUE   FALSE

5. Igneous Rocks are less stable deep with-
    in the earth than Sedimentary Rocks.   TRUE FALSE

6. A rock composed of nearly 100% calcite is

    _____ and given the name _____.

7. At a glance, what is the difference in appearance
    of most rocks and minerals?

    _____

    _____

    _____

8. Weathering and erosion is involved in
    the formation of ? rocks.   _____

9. The most common rock type for the entire
    earth is: _____

10. What are the 1st. order or "primary" rocks? _____

EXERCISE 11-C
INTRODUCTION TO ROCKS AND ROCK CYCLE

Student Name: _____   Lab. Sec. # _____

- -

1. Minerals are composed of elements,
    and rocks are composed of minerals.   TRUE   FALSE

2. Regardless of where found, a rock of
    one particular type (i.e. granite)
    will look <u>exactly</u> like all other rock
    of that type.   TRUE   FALSE

3. The appearance (structure) of a rock
    must be considered in determining its
    classification.   TRUE   FALSE

4. There is more Igneous than Metamorphic
    Rock on earth.   TRUE   FALSE

5. All Igneous Rock on earth is older than
    all Sedimentary Rock.   TRUE   FALSE

6. Metamorphic Rock can form from Igneous
    but not Sedimentary Rock.   TRUE   FALSE

7. Igneous Rock forms only inside the
    earth; only Sedimentary Rock forms on
    the surface.   TRUE   FALSE

8. What is the composition of the mono-
    mineralic rock Syenite?   _____

9. How does a typical rock differ in appearance
    from a monomineralic rock?

    _____

    _____

10. Melting and crystallization are important
    events in the formation of ? rocks.   _____

269

EXERCISE 11-D
INTRODUCTION TO ROCKS AND ROCK CYCLE

Student Name: _____   Lab. Sec. # _____

- -

1. Nearly all rocks are composed of
    minerals and elements.   TRUE   FALSE

2. There is little or no variation in
    the appearance of rock of a particular
    type; if you've seen 1 piece of granite
    you've seen it all.   TRUE   FALSE

3. Both the method of formation and
    mineral composition must be known to
    correctly classify a rock.   TRUE   FALSE

4. One can find Sedimentary and Metamorphic
    Rocks on the earth's surface, but not
    Igneous.   TRUE   FALSE

5. All Sedimentary Rock is cemented together.   TRUE   FALSE

6. A little heat and pressure is likely to
    produce Metamorphic Rock; a lot of heat
    and pressure is likely to produce magma/
    lava thus Igneous Rock.   TRUE   FALSE

7. If all heat and movement within the
    earth were to be lost and stopped
    forever, the surface of earth could
    eventually be all Sedimentary Rock.   TRUE   FALSE

8. What is the composition of the mono-
    mineralic rock Sandstone?   _____

9. What general appearance do all monomineralic
    rocks share?

    _____

    _____

10. Weathering and lithification are
    processes necessary to the formation
    of ? rocks.   _____

271

EXERCISE 11-E
INTRODUCTION TO ROCKS AND ROCK CYCLE

Student Name: _____ Lab. Sec. # _____

- -

1. All rocks contain some mineral
    material.   TRUE   FALSE

2. There can be some variation in the
    appearance of both minerals and
    rocks; all quartz does not look
    exactly the same and neither does
    all granite.   TRUE   FALSE

3. Rocks can be composed of no minerals,
    one mineral, or many minerals. (Hint:
    remember 3 origins for Sedimentary Rocks).   TRUE   FALSE

4. The origin of a rock is of little
    importance in determining its
    classification.   TRUE   FALSE

5. Compaction or dessication are steps
    that might be involved in the formation
    of ___ rocks.   _____

6. You are more likely to encounter
    Sedimentary Rocks than any other type
    of rock on the earth's surface.   TRUE   FALSE

7. Metamorphic rocks are weak and
    unstable in all environments.   TRUE   FALSE

8. Any rock type on the Rock Cycle can
    form directly from any other rock type.   TRUE   FALSE

9. Only Igneous Rock can form directly
    from magma or lava.   TRUE   FALSE

10. If a rock is all quartz, it could be
    the monomineralic rock _____ or
    _____.

EXERCISE 11-F
INTRODUCTION TO ROCKS AND ROCK CYCLE

Student Name: _____   Lab. Sec. # _____

- -

1. The combining of 2 or more minerals
   forms a rock.   TRUE FALSE

2. The properties of rocks are influenced
   by the minerals within them.   TRUE   FALSE

3. If a rock is all calcite it must be
   limestone.   TRUE   FALSE

4. Monomineralic rocks have a different
   appearance than other rocks.   TRUE   FALSE

8. It is possible to recognize minerals
   in some rocks.   TRUE   FALSE

5. What determines whether a rock is
   Igneous, Sedimentary, or Metamorphic?

   _____

   _____

6. Chemical alteration of a rock without
   melting or weathering would produce
   a ? rock.   _____

7. Early in the earth's history, the entire
   planet was in a molten state;
   consequently, there was a time that all
   earth rocks were: _____.

9. Minerals collecting on the ocean floor
   from the evaporation of sea water would
   form ? rock.   _____

10. In your own words, what is the difference
    in rocks and minerals?

    _____

    _____

    _____

    _____

EARTH SCIENCE   LAB. EXERCISE #12
"ROCK CLASSIFICATION AND IGNEOUS ROCKS"

- -

OBJECTIVES:

Be Able to:

1. DISTINGUISH BETWEEN INTRUSIVE AND
    EXTRUSIVE IGNEOUS ROCKS.

2. DISTINGUISH BETWEEN CLASTIC & NONCLASTIC
    SEDIMENTARY ROCKS.

3. DISTINGUISH BETWEEN FOLIATED AND NON-
    FOLIATED METAMORPHIC ROCKS.

4. CORRECTLY IDENTIFY THE NINE IGNEOUS ROCKS
    COVERED IN THIS EXERCISE IF GIVEN
    UNKNOWNS.

5. EXPLAIN THE METHOD OF FORMATION FOR EACH
    OF THE NINE IGNEOUS ROCKS PRESENTED IN
    THIS EXERCISE.

6. DEFINE THE GLOSSARY TERMS.

B.C. by permission of Johnny Hart and Field Enterprises, Inc.

277

GLOSSARY TERMS:

Rock Texture:

Intrusive:

Extrusive:

Clastic:

Nonclastic:

Foliated:

Nonfoliated:

Fine Textured:

Coarse Textured:

Essential Minerals:

Accessory Minerals:

Salt & Pepper Rock:

Monomineralic Igneous Rock:

Vesicles:

Glassy Texture:

Volcanic Bomb:

Bread Bomb:

Cinder-Rock:

Nature's Glass:

Quenched Lava:

Surface Equivalents:

Granite-Rhyolite Family:

Gabbro-Basalt Family:

Isotropic Orientation:

EXERCISE #12
ROCK CLASSIFICATION AND IGNEOUS ROCKS

Recall that rocks are classified into the three families of the "Rock Cycle" (Igneous, Sedimentary, and Metamorphic) based largely on their origin; however, additional subdivision of rocks in these three families is possible based on their individual mineral compositions and structural features. Recognition of differences in composition and appearance enables the naming of specific, individual rock types.

In studying the "Rock Cycle", you learned the following associations:

A. **Igneous Rocks** form from the crystallization of a magma or lava;

B. **Sedimentary Rocks** form from the collection of sediments, precipitated minerals, or organic debris;

C. **Metamorphic Rocks** form from the alteration of previously formed rock without a new cycle of weathering or melting occurring.

We will next discuss the subdivision of each of the three major rock families into their respective sub-groups based on compositional and/or visible structural differences:

Igneous Rocks: can be separated into 2 sub-groups:

A. **Intrustive** = (formed well inside earth)
B. **Extrusive** = (formed on or near earth's surface)

Sedimentary Rocks: can be separated into 2 sub-groups:

A. **Clastic:** (formed from weathered sediments)
B. **Nonclastic:** (formed from mineral ↓ **or** organics)

Metamorphic Rocks: can be separated into 2 sub-groups:

A. **Foliated:** (show layering, platy-minerals, or bands)
B. **Nonfoliated:** (lack folication thus have a uniform look)

INTRUSIVE VS. EXTRUSIVE IGNEOUS ROCKS

Examine some kit specimens to see these visible structural differences. For example, compare the intrusive igneous rock granite, #21 with the extrusive igneous rock rhyolite, #28. In what way are they similar? _____

_____

In what way are they different? _____

_____

The reason for this visual difference is the "rate of cooling" each experienced. When molten rock occurs on or near the earth's surface, it looses its heat quickly; consequently, there is very little time for crystals to grow before the lava is crystallized (solidified). The resulting rock (extrusive) is referred to as a "Fine Textured" rock because you cannot see the individual minerals of which the rock is composed (NOTE: if you can barely see some of the mineral grains it is still alright to refer to the rock as fine textured). Remember: texture in igneous rocks refers to the size of minerals within the rock -- it has nothing to do with the "feel" of the rock. The coarser the texture of an igneous rock, the larger the mineral grains within it; thus, the rock cooled slowly over a long period of time (perhaps hundreds to thousands of years well below the earth's surface (intrusive), allowing crystals to grow quite large. Conversely, fine textured igneous rocks must have formed on or near the earth's surface where the rapid loss of heat did not allow sufficient time for mineral growth to reach sizes easily visible to the naked eye. Examine the nine igneous rocks in your kit (#'s 21-29). Which of these specimens clearly has the finest texture? How did this specimen form?

_____

_____

In your own words, state the relationship between cooling rate, location, and resulting igneous rock texture:

_____

_____

_____

_____

_____

CLASTIC VS. NONCLASTIC SEDIMENTARY ROCKS

Examine some kit specimens to see these visible differences.
For example, compare the clastic sedimentary rock
conglomerate, #32 with the nonclastic sedimentary rock
limestone #33. Describe the visual difference in these two
rocks.

_____

_____

_____

Next compare rock #31 with rock #34. Describe their
differences:

_____

_____

_____

Nonclastic sedimentary rocks therefore have a dense, uniform
look. They appear to be a solid interconnected mass.
Clastic rocks, on the other hand, contain "sediments" that
have been pressed or cemented ("glued") together to form
rock. We will discuss and examine these in more detail in
the next exercise!

- -

FOLIATED VS. NONFOLIATED METAMORPHIC ROCKS

Examine some kit specimens to see these visible differences.
For example, compare the foliated metamorphic rock schist,
#37 with the nonfoliated metamorphic rock quartzite, #39.
Describe the visual difference in these two rocks:

_____

_____

_____

Recheck the definition of foliation. How many of the
metamorphic rocks in the kit (rocks #35 - #40) are foliated?
_____ What do all nonfoliated metamorphic rocks have in
common? (HINT: composition).

_____

We will discuss and examine these in more detail in the next
exercise!

For the remainder of this exercise will focus on the properties of the <u>Igneous</u> <u>Rocks</u> and properties which enable their identification.

INTRUSIVE IGNEOUS ROCKS -- (Coarse Textured)

#21. GRANITE

"Essential minerals" are quartz and feldspar; contains much quartz; white to pink minerals with cleavage surfaces are feldspars; "accessory minerals" are typically micas and hornblende; this is the most common rock in the continental portion of the earth's crust. <u>Why</u> <u>do</u> <u>you</u> <u>think</u> <u>this</u> <u>is</u> <u>such</u> <u>a</u> <u>popular</u> <u>building</u> <u>material?</u> _____

_____.

What advantage do man-made building materials have over a natural material like granite?

_____

NOTES: <u>hornblend, feldspar (Gray or pink), quartz</u>

<u>Intrusive, Igneous</u>

_____

#22 DIORITE

The major difference in diorite and granite is the lack or scarcity of quartz in diorite compared to the quartz-rich granite; otherwise they are very similiar; diorite is often called the "salt and pepper rock" because of its 50-50 mix of feldspar (light colors) and hornblende (dark colors). <u>How</u> <u>do</u> <u>you</u> <u>suppose</u> <u>diorite</u> <u>got</u> <u>its</u> <u>name?</u> _____

_____

How can the presence or absence of quartz in a rock be determined so you can distinguish between granite and diorite? _____

_____

NOTES: <u>50:50 feldspar + hornblend</u>

<u>Intrusive, Igneous</u>

_____

22,23 & 24
Same composion as granite,
but no quartz

#23. SYENITE

This is a "monomineralic" (1 mineral) igneous rock composed of the mineral feldspar; at first glance the rock may not appear to be coarse textured but look at it carefully! When turned in the light you can see the individual crystal surfaces reflecting light. Compare this look with specimen #28 which is "fine textured". If this rock had quartz in it what would it be called? _____

NOTES: almost all ~~hornblend~~ feldspar, with ~~lesser~~ hornblend

Intrusive, Igneous Rock

_____

#24. GABBRO

Like syenite, you need to examine this intrusive rock closely to see that it has a coarse texture; it is composed totally of dark minerals, like hornblende, so there is little or no color contrast; the rock will always be black to dark green and coarser in texture than any similiar rocks; furthermore, the crystals are intergrown in an even, mosaic pattern. Do you think that some gabbro could be monomineralic? _____
If gabbro had large amounts of light colored feldspars within it what would it be called? _____

NOTES: almost all ~~hornblend~~ Hornsblend, w/ some feldspar

Intrusive, Igneous Rock.

_____

==

EXTRUSIVE IGNEOUS ROCKS -- (Fine Textured)

#25. PUMICE

An unusual, and easy to identify rock; pumice is always full of holes (called "vesicles") formed when trapped gases exploded from the lava as it was thrown into the atmosphere during a volcanic eruption; the solid rock itself cooled so fast that it has a glassy (extremely fine) texture; the rock is typically so light that even large pieces float in water; most often light colors (grays, white, tan, etc); Examine a specimen of pumice with some magnification to fully appreciate its fine texture. Why do you think many pieces of pumice are streamlined like a football? _____

Some pumice is referred to as "bread bombs"--what might be
the origin of that name? _____

NOTES: _____

_____

_____

#26. SCORIA

Scoria is somewhat similiar in appearance to pumice but is
always darker and heavier; this is due to the presence of
larger amounts of iron and heavy elements; the rock typically
looks as if it has been scorched by heat; to some people the
rock is reminiscent of the "cinders" or "clinkers" removed
from old coal furnaces; notice that scoria is not as fine
textured as pumice and, also, scoria will not float; most
scoria forms on the top of lava flows where gases are still
escaping as the rock is forming. What would be indicated if
vesicles were long and narrow instead of round? What does
the amount of scoria in an area indicate about the lava it
formed from? _____

_____

NOTES: _Red Volcanic Sender, denser then pumice_
_____extrusive, Igneous rock_____

_____

#27. OBSIDIAN

This unusual rock is often referred to as "nature's glass";
it is only formed when extremely hot volcanic lava flows
quickly into water; such an action is said to "quench" a
lava; obsidian will shatter like glass and often displays
"conchoidal" fractures as seen in quartz; primitive people
were fascinated by obsidian and used/traded it alot; thin
pieces of obsidian are used in delicate micro-surgery as
scaples. -- Why do you think it has that use?

natures glass, Jet black concodeal fracture
extrusive, Igneous Rock

Do you think there is much range in composition of obsidian?
_____ Why? _____

_____

Can any/all lava flowing into water produce obsidian? _____

Why? _____

Extrusive igneous rock makes-up most of the ocean sea floor rock, yet obsidian is not very common -- Why?

_____

_____

NOTES: _____

_____

_____

#28. RHYOLITE

This rock is known as the "surface equivalent" of granite; together they form what is referred to as the "granite-rhyolite family"; these 2 rocks have identical mineral compositions, but greatly different textures because of their location, thus rate, of formation; melt rhyolite and allow it to cool slowly and granite is formed; melt rhyolite and allow it to cool quickly and rhyolite is formed; rhyolite is nearly always a pinkish-gray color and usually appears to be banded or layered; this is the most common rock formed by lava flows on the earth's landmasses; like granite the rock is mostly feldspar and quartz. What color are the different layers of rhyolite? _____. What do these different colored layers represent? _____
Some people confuse rhyolite with the rock sandstone #31; examine these together! How can they be distinguished from each other? _____

_____

NOTES: _PinKish color, extrusive, Igneous Rock._

_____

_____

#29. BASALT

Basalt is known as the "surface equivalent" of gabbro; together they form what is referred to as the "gabbro-basalt family"; these 2 rocks also have identical mineral compositions, but greatly different textures for the same reasons seen in granite-rhyolite; basalt is always dark, and quite dense; it can have some vesicles but they are not very common; if a dark igneous rock has a great amount of vesicles, call it scoria!

Basalt is the most common rock formed by lava flows on the earth's ocean basin floors; like gabbro, the rock is mostly hornblende and related dark, heavy minerals. <u>Why do the earth's landmasses (continents) sit passively upon the rocks of the ocean basin floors?</u>
<u>Why might it be difficult to distinguish between basalt and gabbro in some cases?</u> _____
_____

Is it possible to form obsidian from the same lava that forms gabbro? _____ If yes, How?_____

_____
NOTES: Black dull color, very common lava flow rock, extrusive, Igneous Rock
_____

IN SUMMARY:

┌─────────────────────────────────────┐
│ CHARACTERISTICS OF IGNEOUS ROCKS │
└─────────────────────────────────────┘

A. HARD, STRONG ROCKS WITH HIGH S.G.
    (UNLESS HIGHLY VESICULAR);

B. IRREGULAR SHAPED INDIVIDUAL MINERAL
    GRAINS EVERYWHERE IN CONTACT WITH
    ALL OTHER MINERAL GRAINS
    (NOT VISIBLE IN EXTRUSIVES);

C. LACK OF VOID SPACE (EXCEPT VESICULAR
                        ROCKS);

D. MINERAL GRAINS ARE EQUIDIMENSIONAL
    IN ALL DIRECTIONS -- THEY ARE NOT
    ORIENTED IN ANY SPECIFIC DIRECTION
    (REFERRED TO AS "<u>ISOTROPIC</u>").

┌──────────────────────┐
│ THOUGHT QUESTIONS: │
└──────────────────────┘

1. Igneous Rocks make up 95% of the outer 10 miles of the earth and nearly 100% below a depth of 10 miles. Why are they so common?

    molten mAterial
_____

Why are they less common on the continental land surfaces?

because of weathering

2. Do you think the chemistry (mineral composition) of Igneous Rocks is more <u>or</u> less complex than Sedimentary or Metamorphic Rocks? _Less_ Why?

they have more sediments in them

3. Which Igneous Rocks chemically decompose (break down) fastest on the earth's surface?

Scoria, pumice

Why? _because of their vesicles,_

4. Can good soils form from the weathering of igneous rocks? _____ If yes, which produce the best soils?

yes pumice, ash

Why? _____

NOTE: BE SURE TO EXAMINE A VARIETY OF IGNEOUS ROCKS TO SEE A VARIETY OF "LOOKS" THE INDIVIDUAL TYPES CAN DISPLAY!

EXERCISE 12-A
ROCK CLASSIFICATION AND IGNEOUS ROCKS

Student Name: _____ Lab. Sec. # _____

- -

1. Intrusive igneous rocks contain minerals
    large enough to be identified.   TRUE   FALSE

2. Rocks formed from the weathered debris
    of other rocks are called ? Sedimentary
    Rocks.   _____

3. The finest textured igneous rock is: _____

4. A quartz-like, intrusive ingeous rock
    would be given the name: _____

==
Using the set of rocks you have been given, identify the
unknowns and their texture.
==

ROCK NAME	COARSE OR FINE TEXTURE
5. _____	_____
6. _____	_____
7. _____	_____
8. _____	_____
9. _____	_____
10. _____	_____

BONUS: _____   _____

__Record your unknown set #__ _____

EXERCISE 12-B
ROCK CLASSIFICATION AND IGNEOUS ROCKS

Student Name: _____   Lab. Sec. # _____

- -

1. Individual minerals in extrusive
    igneous rocks are too small to be
    identified.   TRUE   FALSE

2. Rocks formed from the compaction of
    organic materials are called ?
    Sedimentary Rocks.   _____

3. The most "vesicular" igneous rock is: _____

4. The only difference in basalt and
    gabbro is: _____

==
Using the set of rocks you have been given, identify the
unknowns and their texture.
==

ROCK NAME	COARSE OR FINE TEXTURE
5. _____	_____
6 _____	_____
7. _____	_____
8. _____	_____
9. _____	_____
10. _____	_____

BONUS: _____   _____

Record your unknown set # _____

EXERCISE 12-C
ROCK CLASSIFICATION AND IGNEOUS ROCKS

Student Name: _____ Lab. Sec. # _____

- -

1. The faster a lava or magna cools,
    the coarser the resulting rock
    texture.   TRUE   FALSE

2. The collection of minerals on the
    sea floor because of the evaporation
    of water can form ? Sedimentary Rocks.   _____

3. The most common igneous rock on the
    world's ocean floors is: _____

4. A major difference in diorite and
    gabbro is: _____

===
Using the set of rocks you have been given, identify the
unknowns and their texture.
===

ROCK NAME   **COARSE OR FINE TEXTURE**

5. _____   _____

6   _____   _____

7. _____   _____

8. _____   _____

9. _____   _____

10. _____   _____

BONUS: _____   _____

Record your unknown set # _____

EXERCISE 12-D
Rock Classification And Igneous Rocks

Student Name: _____   Lab. Sec. # _____

- -

1. Rock texture in igneous rocks is more
    a result of rate of cooling than
    mineral composition.   TRUE   FALSE

2. If Metamorphic Rocks display layers or
    bands they are said to be: _____

3. The most common extrusive igneous rock
    on the earth's continents is: _____

4. The only igneous rock that could have
    the same composition as any or all other
    igneous rocks is: _____

==
Using the set of rocks you have been given, identify the
unknowns and their texture.
==

ROCK NAME   **COARSE OR FINE TEXTURE**

5. _____   _____

6   _____   _____

7. _____   _____

8. _____   _____

9. _____   _____

10. _____   _____

BONUS: _____   _____

Record your unknown set # _____

EXERCISE 12-E
ROCK CLASSIFICATION AND IGNEOUS ROCKS

Student Name: _____ Lab. Sec. # ____

- -

1. Lavas typically produce coarser
   textures than magnas.   TRUE   FALSE

2. A uniform, dense-looking Metamorphic
   Rock would be classified as: _____

3. A dark, highly vesicular igneous rock
   would be given the name: _____

4. It is possible to find coarse textured
   igneous rocks on the earth's surface.   TRUE   FALSE

==
Using the set of rocks you have been given, identify the
unknowns and their texture.
==

ROCK NAME	COARSE OR FINE TEXTURE
5. _____	_____
6 _____	_____
7. _____	_____
8. _____	_____
9. _____	_____
10. _____	_____

BONUS: _____   _____

Record your unknown set # _____

EXERCISE 12-F
ROCK CLASSIFICATION AND IGNEOUS ROCKS

Student Name: _____   Lab. Sec. # _____

- -

1. It can be difficult or impossible to classify some rocks are clearly intrusive or extrusive.   TRUE   FALSE

2. All igneous rocks formed well below the earth's surface are classified as: _____

3. A given magma might cool to form granite, or, flow onto the surface as a lava and form the rock basalt.   TRUE   FALSE

4. The 2 most common igneous rocks formed

    extrusively are _____ and

    _____ .

==
Using the set of rocks you have been given, identify the unknowns and their texture.
==

	ROCK NAME	COARSE OR FINE TEXTURE
5.	_____	_____
6	_____	_____
7.	_____	_____
8.	_____	_____
9.	_____	_____
10.	_____	_____

BONUS: _____   _____

Record your unknown set # _____

EARTH SCIENCE   LAB. EXERCISE #13

"SEDIMENTARY AND METAMORPHIC ROCKS"

- -

OBJECTIVES:

Be Able to:

1. DISTINGUISH BETWEEN CLASTIC AND NON-
    CLASTIC SEDIMENTARY ROCKS.

2. DISTINGUISH BETWEEN FOLIATED AND NON-
    FOLIATED METAMORPHIC ROCKS.

3. CORRECTLY IDENTIFY THE 5 SEDIMENTARY
    AND 6 METAMORPHIC ROCKS COVERED IN
    THIS EXERCISE IF GIVEN UNKNOWNS.

4. EXPLAIN THE METHOD OF FORMATION FOR
    EACH OF THE 11 ROCKS COVERED IN THIS
    EXERCISE.

5. DEFINE THE GLOSSARY TERMS.

B.C. by permission of Johnny Hart and Field Enterprises, Inc.

GLOSSARY TERMS:

Fissile:

Cross-bedding:

Friable:

Matrix:

Sorting:

Breccia:

Wentworth Scale:

Dimension Stone:

Low Grade Coal:

High Grade Coal:

Peat:

Lignite:

Bituminous:

Anthracite:

Hard Coal:

Soft Coal:

Fossil Fuel:

Low Grade Metamorphism:

High Grade Metamorphism:

Foliation:

Platy Minerals:

Parent Rock:

High Grade Marble:

Low Grade Marble:

"Indiana Marble":

Metamorphic Equivalents:

EXERCISE #13
SEDIMENTARY AND METAMORPHIC ROCKS

In this final laboratory exercise you will learn to identify the more common Sedimentary and Metamorphic Rocks. It is important that you recall, and review if necessary, the basic characteristics of both Sedimentary and Metamorphic Rocks. Remember that Sedimentary Rocks are classified into 2 groups (Clastic and Nonclastic), as are the Metamorphic Rocks (Foliated and Nonfoliated).

```
SEDIMENTARY ROCKS
```

CLASTICS

#30. SHALE

Composed of the smallest sediments (clay); color will vary greatly (red, green, brown, etc.) but is most commonly black or gray; often occurs in thin layers which break into thin sheets (Fissile); can also be massive and dense looking; usually very soft (often can be scratched with thumbnail); becomes "muddy" when wetted; emits "musty odor" when damp because of clay minerals (Kaolinite); most common rock on earth's surface. Why is shale the most common rock on earth?

_____

_____

Of what economic importance is shale? _____

_____

How does shale form? _____

_____

NOTES: _Fissle (flat-layers), musty odor, dull black or gray,_ ____

_____

_____

#31. SANDSTONE

Composed of sand-sized sediments; gritty feel; usually will scratch glass because of sand-sized quartz grains; often layered; sometimes shows "cross-bedding"; many colors possible from white to browns, to reds; second most common rock on earth's surface. Why is quartz the dominant mineral in sandstone? _____

How does sandstone form? _____

_____

Some sandstone is so weak that the sand grains can easily be rubbed loose ("Friable"). Why are some sandstones weak and others very hard? _____
What is the most common cement ("matrix") for sandstone? _____

NOTES: ~~sandpaper feel~~, beeding Horizontal lines _____

_____

_____

#32.   CONGLOMERATE

Composed of any combination of sediments of which the majority are larger than sand; typically a "poorly sorted" array of pebbles and granules which are well worn (rounded) due to weathering; looks as if someone dropped cement in a gravel driveway to form the rock; if the sediments are sharp and angular the rock is referred to as "breccia"; least common sedimentary rock. Why is conglomerate so uncommon?

_____

How does conglomerate form? _____

_____

What is the difference in environments where shale and conglomerate form? _____

_____

NOTES: verity of different rocks & sizes _____

_____

_____

- -

WENTWORTH SCALE

SEDIMENT SIZES

SEDIMENT	DIAMETER	IN MM	ROCK NAME
BOULDERS..........> 256	...........Conglomerate/Breccia
COBBLES..............256	- 64.......Conglomerate/Breccia
PEBBLES...............64	- 4.......Conglomerate/Breccia
GRANULES...............4	- 2........Conglomerate/Breccia
SAND....................2	- 1/16....Sand
SILT............ .1/16	- 1/256...Siltstone
CLAY............< 1/256	- ....Shale

know

soil

which

#33. LIMESTONE (Chemically formed)

A monomineralic rock composed entirely of calcite deposited as a ↓ from evaporating seawater; rock has a fast, violent reaction with HCl; many colors are possible but light grays and tans are most common; fossils are also common in many limestones; rock is used as a building material ("dimension stone"), and is crushed into powder for "Ag Lime" to treat ("sweeten") the pH of fields/lawns; the 3rd. most common sedimentary rock on the earth's surface; presence of calcite can give the rock an igneous-like crystalline look; only has a hardness of 3. What are the advantages to using limestone as a building material? _____

_____

What are some disadvantages? _____

_____

Why is Indiana so famous for limestone? _____

_____

How does limestone form? _____

_____

NOTES: Calcium, HCl Test _____

_____

_____

#34. COAL (Organically formed)

Coal is a very light (low S.G.), black, shiny rock; easily crushed, it typically breaks with a "blocky fracture"; names for coal depend upon its grade (amount of B.T.U.'s of heat produced); grade is a result of the amount of pressure involved in the formation of coal; from lowest to highest the grades of coal are: PEAT, LIGNITE (Brown), BITUMINOUS (soft), and ANTHRACITE (Hard); hard coal or anthracite is considered a metamorphic rock; coal is called a "fossil fuel" because it formed from the organic remains of pre-historic (fossil) plants. What advantages do higher grades of coal have over lower grades? _____

_____

What is the problem with much of the coal in this part of the country? _____

How does "hard coal" form? _____

_____

What rock might be confused with coal? _____
How can they be distinguished? _____

_____

NOTES: *Black, shiny, lightweight* _____

_____

_____

```
SUMMARY OF SEDIMENTARY ROCKS
```

```
1.)   TYPICALLY LAYERED & OCCUR IN BEDS;
2.)   SOFT, WEAK ROCKS WITH A LOW S.G.;
3.)   FOSSILS OFTEN PRESENT;
4.)   COARSE CLASTICS HAVE WORN GRAINS;
5.)   VOID SPACES COMMON IN CLASTICS;
6.)   FORM IN A WIDE RANGE OF ENVIRONMENTS;
7.)   WIDE VARIETY OF COMPOSITIONS & TEXTURES;
8.)   MAKE UP 75% OF THE CONTINENTAL LANDMASS SURFACE;
9.)   MAKE UP ONLY 5% OF THE OUTER 10 MILES OF EARTH;
```

- -

```
METAMORPHIC ROCKS
```

FOLIATED

#35. SLATE

A low grade metamorphic rock (forms with little heat/pressure); the "metamorphic equivalent of shale"; very similar in appearance to shale but is harder, has a brittle "ring" when tapped, layers are thinner and more tightly packed, surface may display a "shine"; if equal in size, the slate will be heavier than shale; many colors possible but black is most common; splits into very even, thin sheets.
Of what economic importance is slate? _____

_____

What properties of slate make it useful? _____

_____

How does slate form? _____

_____

NOTES: _____

_____

_____

#36. PHYLLITE

A low to intermediate grade metamorphic rock, but higher
grade than slate; most commonly a dull brown to greenish
color; layers are thicker and less uniform (often wavy) than
slate; surface is very distinct - a satin-like shine with a
rippled surface that reminds many of "snake skin"; name is
pronounced "Fill-ite". Are slate and phyllite likely to be
found in the same localities? _____ Why? _____

_____

Why is phyllite classified as foliated? _____

NOTES: Satin luster, slight different color _____

_____

_____

#37. SCHIST

An intermediate grade metamorphic rock; composed of platy
(flakelike) minerals such as mica and talc; a variety of
minerals can frequently be identified with the naked eye;
layers are often present but are wavy or contorted; forms
from a great variety of "parent rocks"; named for the most
abundant or conspicuous mineral (i.e. garnet-schist,
mica-schist, hornblende-schist etc.); weathers quickly when
exposed to the atmosphere.
Why is schist considered a foliated metamorphic rock?
_____

Why does schist weather so quickly and easily? _____

Is schist more likely to occur with slate, or phyllite?
_____ Why? _____

NOTES: Shinny, _____

_____

_____

#38. GNEISS

Pronounced "nice"; a high grade metamorphic rock; the "metamorphic equivalent of granite"; characterized by bands or zones of minerals of different colors that may be parallel or highly twisted throughout the rock; compare a piece of gneiss with a piece of granite; how are they different?

_____

_____

How do the bands or zones of minerals form within the gneiss?

_____

_____

How close did gneiss come to being an igneous rock?

_____

Why is gneiss classified as foliated? _____

_____

NOTES: _____

_____

_____

_____

- -

NONFOLIATED

#39. QUARTZITE

A high grade, monomineralic metamorphic rock; forms at such high temperatures that sand quartz grains and the quartz cement holding them together become fused into one solid mass of rock; most typical are colors of white, light browns, greens, or reds; a rather smooth, featureless rock, it can display conchordal fractures when broken; extremely hard and resistant to erosion; the "metamorphic equivalent of quartz sandstone". How can quartzite be distinguished from sandstone?

Why isn't quartzite used as a building material? Why is there less quartzite than slate or gneiss? _____

Why is quartzite classified as "nonfoliated"? _____

NOTES: Earth tone Colors _____

_____

_____

#40. MARBLE

A low to high grade monomineralic metamorphic rock; it is the "metamorphic equivalent" of limestone; principal mineral is calcite thus rock will react quickly and violently to HCl; if a weak and slow reaction occurs, dolomite may be the dominant mineral; light pinks and white with swirls of other colors are common; low grade marble has a fine texture (smooth) and is therefore favored for art work; high grade marble was the result of much heat/pressure which allowed crystals to grow much larger (rough) coarse textured; in low grade marble one can sometimes find leftover features from the parent limestone; such features are called "relic structures".

Why is marble a common rock used by sculpturers?

_____

_____

Why is marble used as a decorative rock in expensive homes and buildings? _____

_____

Why is marble a poor choice for a "tombstone"? _____

_____

How does marble form? _____

_____

There is no marble in Indiana, but there is rock referred to as "Indiana Marble". What do you suppose this rock actually is? _____

NOTES: _used to be limestone Hcl test   gray or white_

_____

_____

309

SUMMARY OF METAMORPHIC ROCKS

1.) CONSIST OF BANDED ZONES OF MINERALS; OR
2.) PLATY MINERALS OF SIMILAR ORIENTATION; OR
3.) LAYERS THROUGHOUT THE ROCK.
4.) IF NOT FOLIATED, ARE MONOMINERALIC.
5.) LEAST COMMON OF THE 3 ROCK GROUPS.
6.) MOST ROCKS ARE STRONG WITH HIGH S.G.
7.) ALL ARE THE RESULT OF ALTERATION OF OTHER ROCK.

METAMORPHIC EQUIVALENTS

PARENT ROCK	METAMORPHIC EQUIVALENT
SHALE	SLATE
SHALE OR SLATE	PHYLLITE
SHALE, SLATE, PHYLLITE	SCHIST
IMPURE SANDSTONE	SCHIST
IMPURE LIMESTONE	SCHIST
BASALT	SCHIST
SCHIST	GNEISS
IMPURE SANDSTONE	GNEISS
SHALE	GNEISS
GRANITE - GABBRO	GNEISS
SANDSTONE	QUARTZITE
LIMESTONE	MARBLE

NOTE: BE SURE TO EXAMINE A NUMBER OF SEDIMENTARY
AND METAMORPHIC ROCKS TO SEE A VARIETY OF
"LOOKS" THE INDIVIDUAL TYPES CAN DISPLAY.

EXERCISE 13-C
SEDIMENTARY & METAMORPHIC ROCKS

Student Name: _____ Lab. Sec. # _____

- -

Using the set of rocks you have been given, identify the unknowns and their sub-group (i.e. clastic or nonclastic; foliated or nonfoliated).

- -

ROCK NAME   SUB-GROUP

1. _____   _____

2. _____   _____

3. _____   _____

4. _____   _____

5. _____   _____

6. _____   _____

7. _____   _____

8. _____   _____

9. _____   _____

10. _____   _____

Record you unknown set # _____

EXERCISE 13-D
SEDIMENTARY & METAMORPHIC ROCKS

Student Name: _____ Lab. Sec. # _____

- -

Using the set of rocks you have been given, identify the unknowns and their sub-group (i.e. clastic or nonclastic; foliated or nonfoliated).

- -

ROCK NAME   SUB-GROUP

1. _____   _____

2. _____   _____

3. _____   _____

4. _____   _____

5. _____   _____

6. _____   _____

7. _____   _____

8. _____   _____

9. _____   _____

10. _____   _____

Record you unknown set # _____

EXERCISE 13-E
SEDIMENTARY & METAMORPHIC ROCKS

Student Name: _____ Lab. Sec. # _____

- -

Using the set of rocks you have been given, identify the unknowns and their sub-group (i.e. clastic <u>or</u> nonclastic; foliated or nonfoliated).

- -

<u>ROCK NAME</u>	<u>SUB-GROUP</u>
1. _____	_____
2. _____	_____
3. _____	_____
4. _____	_____
5. _____	_____
6. _____	_____
7. _____	_____
8. _____	_____
9. _____	_____
10. _____	_____

<u>Record you unknown set #</u> _____

EXERCISE 13-F
SEDIMENTARY & METAMORPHIC ROCKS

Student Name: _____ Lab. Sec. # _____

- -

Using the set of rocks you have been given, identify the unknowns and their sub-group (i.e. clastic or nonclastic; foliated or nonfoliated).

- -

ROCK NAME | SUB-GROUP

1. _____   _____

2. _____   _____

3. _____   _____

4. _____   _____

5. _____   _____

6. _____   _____

7. _____   _____

8. _____   _____

9. _____   _____

10. _____   _____

Record you unknown set # _____

APPENDIX

CONVERSION FACTORS FOR COMMONLY USED UNITS

Length	Å	in.	m	cm
1 ångström (Å)	1	3.94×10^{-9}	10^{-10}	10^{-8}
1 inch (in.)	2.54×10^8	1	2.54×10^{-2}	2.54
1 meter (m)	10^{10}	39.37	1	10^2
1 centimeter (cm)	10^8	.3937	10^{-2}	1

Mass	lb	oz	kg	g
1 pound (lb)	1	16	.4536	453.6
1 ounce (oz)	.0625	1	2.836×10^{-2}	28.36
1 kilogram (kg)	2.204	35.3	1	1000
1 gram (g)	2.204×10^{-3}	.0353	.001	1

Pressure	atm	Pa	mm Hg	dyn cm^{-2}	lb in.$^{-2}$
1 atmosphere (atm)	1	1.013×10^5	760	1.013	14.70
1 pascal (Pa)	9.872×10^{-6}	1	7.502×10^{-3}	10^{-5}	1.451×10^{-4}
1 torr (mm Hg)	1.32×10^{-3}	1.333×10^2	1	1.33×10^{-5}	1.93×10^{-2}
1 bar (dyn cm^{-2})	9.872×10^{-1}	10^5	7.502×10^2	1	1.451×10^1
1 pound per square inch (lb in.$^{-2}$)	6.803×10^{-2}	6.891×10^3	5.17×10^1	6.891×10^{-2}	1

Temperature	°K	°F	°C
1 degree Kelvin (°K)	1	$\frac{9}{5}(°K) - 459.7$	$°K + 273.16$*
1 degree Fahrenheit (°F)	$\frac{5}{9}(°F) + 255.4$	1	$\frac{5}{9}(°F - 32)$
1 degree Centigrade (°C)	$°C - 273$	$\frac{9}{5}(°C) + 32$	1

Energy	J	erg	cal	kWh
1 joule (J)	1	10^7	.2390	2.8×10^{-7}
1 erg (erg)	10^{-7}	1	2.390×10^{-8}	2.8×10^{-14}
1 thermochemical calorie (cal)	4.184	4.184×10^7	1	1.162×10^{-6}
1 kilowatt-hour (kWh)	3.6×10^6	3.6×10^{13}	8.604×10^5	1

Volume	ml	cm^3	qt	oz
1 milliliter (ml)	1	1	1.06×10^{-3}	3.392×10^{-2}
1 cubic centimeter (cm^3)	1	1	1.06×10^{-3}	3.392×10^{-2}
1 quart (qt)	943	943	1	32
1 fluid ounce (oz)	29.5	29.5	3.125×10^{-2}	1

*Absolute zero (°K) = -273.16°C.

SELECTED TABLES OF WEIGHTS AND MEASURES

LINEAR MEASURE

1 inch = 2.54 centrimeters
12 inches = 1 foot = .3048 meter
3 feet = 1 yard = .9144 meter
16.5 feet = 5.5 yards = 1 rod (pole or perch) in 5.029 meters
40 rods = 1 furlong = 201.17 meters
8 furlongs = 5280 feet = 1760 yards = 1 mile = 1609.3 meters
3 miles = 1 league - 4.83 kilometers

CHAIN MEASURE (SURVEYORS CHAIN)

7.92 inches = 1 link = 20.12 centimeters
100 links = 66 feet = 1 chain = 20.12 meters
10 chains = 1 furlong = 201.17 meters
80 chains = 1 mile = 1609.3 meters

SURVEYOR'S MEASURE

625 square links = 1 square pole = 25.29 square meters
16 square poles = 1 square chain = 404.7 meters
10 square chains = 1 acre = .4047 hectare
640 acres =1 sq. mile =1 section = 259 hectares = 2.59 sq. km
36 square miles =1 township =9324 hectares =93.24 sq. km.

NAUTICAL MEASURE

6 feet = 1 fathom = 1.829 meters
100 fathoms = 1 cable's length = 120 fathoms in U.S. Navy!
10 cable's lengths = 1 nautical mile = 6076.1 ft. = 1.852 km.
1 nautical mile = 1.1508 statute miles = 1 ' of longitude at
3 nautrical miles = 1 marine league = 3.45 statute miles
60 nautical miles = 1 degree of a great circle of the earth

LAND MEASURE

1 square meter = 1 centaire = 1549.9 square inches
100 centiares = 1 are = 119.6 square yards
100 ares = 1 hectare = 2.471 acres
100 hectares = 1 square kilometer = .386 square miles

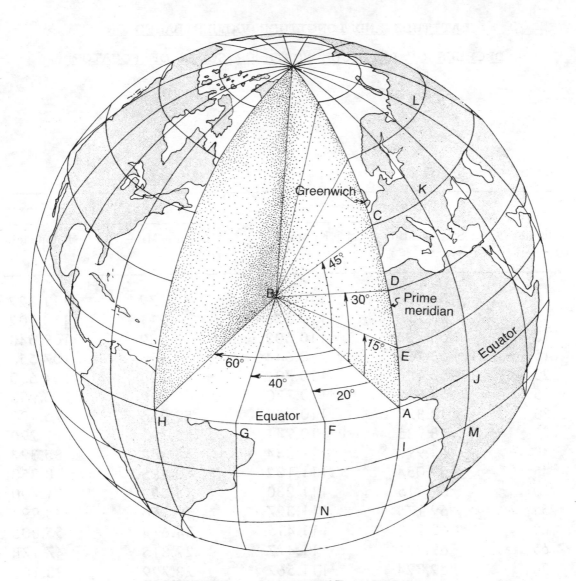

"LATITUDE – LONGITUDE MODEL"

===

| Wind speed (mph) | Actual temperature (°F.) | | | | | | | | | | | |
| | 50 | 40 | 30 | 20 | 10 | 0 | −10 | −20 | −30 | −40 | −50 | −60 |
	Equivalent temperature (°F.)											
Calm	50	40	30	20	10	0	−10	−20	−30	−40	−50	−60
5	48	37	27	16	6	−5	−15	−26	−36	−47	−57	−68
10	40	28	16	4	−9	−21	−33	−46	−58	−70	−83	−95
15	36	22	9	−5	−18	−36	−45	−58	−72	−85	−99	−112
20	32	18	4	−10	−25	−39	−53	−67	−82	−96	−110	−124
25	30	16	0	−15	−29	−44	−59	−74	−88	−104	−118	−133
30	28	13	−2	−18	−33	−48	−63	−79	−94	−109	−125	−140
35	27	11	−4	−20	−35	−49	−67	−82	−98	−113	−129	−145
40	26	10	−6	−21	−37	−53	−69	−85	−100	−116	−132	−148

Wind speeds greater than 40 miles per hour have little additional effect.

"WIND CHILL CHART"

LATITUDE AND LONGITUDE VALUES BASED ON
DEGREES OF LATITUDE NORTH OR SOUTH OF EQUATOR

Latitude (Degrees)	LENGTH OF 1° OF LATITUDE		LENGTH OF 1° OF LONGITUDE	
	Statute Miles	Kilometers	Statute Miles	Kilometers
0	68.704	110.569	69.172	111.322
5	68.710	110.578	68.911	110.902
10	68.725	110.603	68.129	109.643
15	68.751	110.644	66.830	107.553
20	68.786	110.701	65.026	104.650
25	68.829	110.770	62.729	100.953
30	68.879	110.850	59.956	96.490
35	68.935	110.941	56.725	91.290
40	68.993	111.034	53.063	85.397
45	69.054	111.132	48.995	78.850
50	69.115	111.230	44.552	71.700
55	69.175	111.327	39.766	63.997
60	69.230	111.415	34.674	55.803
65	69.281	111.497	29.315	47.178
70	69.324	111.567	23.729	38.188
75	69.360	111.625	17.960	28.904
80	69.386	111.666	12.051	19.394
85	69.402	111.692	6.049	9.735
90	69.407	111.700	0.000	0.000

For Use with Exercise 1   (Latitude & Longitude)

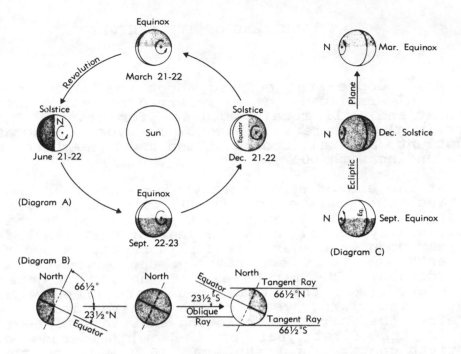

"EARTH-SUN RELATIONSHIPS"

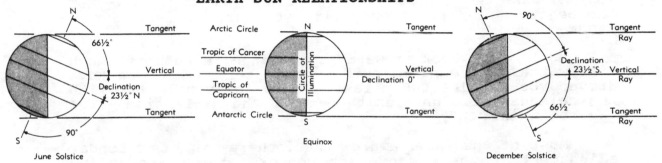

Latitude	March 20–21	June 21–22	September 22–23	December 22–23
0°	12 hr	12 hr	12 hr	12 hr
10°	12 hr	12 hr 35 min	12 hr	11 hr 25 min
20°	12 hr	13 hr 12 min	12 hr	10 hr 48 min
30°	12 hr	13 hr 56 min	12 hr	10 hr 4 min
40°	12 hr	14 hr 52 min	12 hr	9 hr 8 min
50°	12 hr	16 hr 18 min	12 hr	7 hr 42 min
60°	12 hr	18 hr 27 min	12 hr	5 hr 33 min
70°	12 hr	2 months	12 hr	0 hr 0 min
80°	12 hr	4 months	12 hr	0 hr 0 min
90°	12 hr	6 months	12 hr	0 hr 0 min

LENGTH OF DAYLIGHT IN NORTHERN HEMISPHERE

THE END OF THE CENTURY

Most people are confused about the calendar system now in universal use. It is referred to as the Gregorian calendar and will remain accurate for several thousand years in the future. It needs, however, an occasional adjustment, and this occurs at the **end** of each century in years ending with 00.

At the end of each century with each year ending in 00? Yes, the years ending in 00 (even hundreds) signify the _end_ of a century. The first year, the year traditionally looked on as the year of Christ's birth and the year upon which our calendar is based, was year one. The tenth was year ten, and was the last year of the first decade. The first year of the second decade was year eleven, and the second ten years ended with year twenty. The first year of the third decade would, therefore, be year 21. Following that logical beginning, 1981 is the first year of the ninth decade of the twentieth century, and the year 2000 will be the **last** year of the tenth decade of the twentieth century - **NOT** the beginning of the twenty-first century as is commonly assumed.

Most of us look forward to a change in numbers, much like watching the nines turn over to zeros when you have driven you car 100,000 miles. It is an event. But with the calendar it signifies the **end** of the millennium - **NOT** the beginning.

Most of us are aware that there is a calendar adjustment at the end of each century, the years ending in 00. But most people have the adjustment backwards. Every four normal years an extra day is added at February 29, and we are used to that phenomenon as leap year. This is an imperfect adjustment, however. In 100 years the calendar gets a little out of whack and it is adjusted by omitting leap year. at the end of each century, years ending in 00. But even this centennial adjustment is not perfect. So in all years evenly divisible by 400 an exception is made and the leap year remains. Therefore, our leap year cycle is consistent for the years 1904 through the year 2096, and the upcoming end of the millennium, year 2000, **is a leap year.**

The leap year cycle breaks at 1900 and 2100. This indicates the omission of leap year at those times. But note, the leap year cycle does not break at the year 2000, indicating that the year 2000 is a leap year and does continue the four-year leap year sequence.

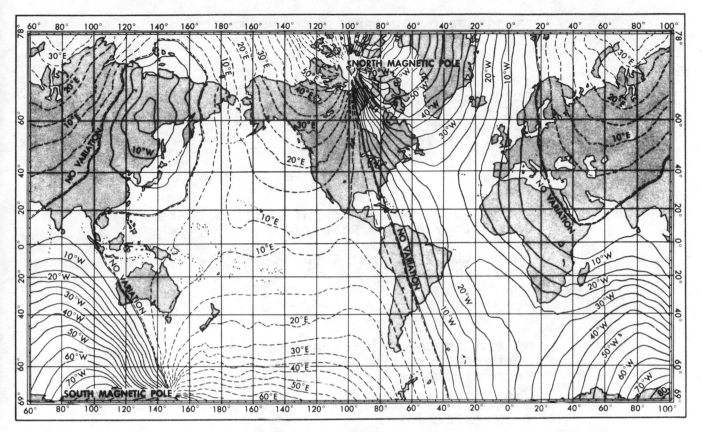

"ISOGONIC LINES"

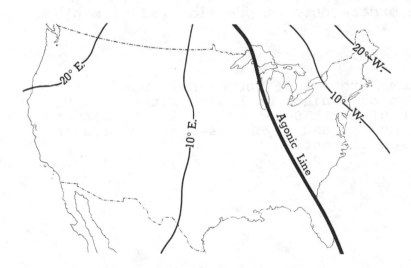

MAGNETIC DECLINATION OF U.S.

TOPOGRAPHIC MAP ORDERING INFORMATION

Indiana Geological Survey
Publication Section
611 N. Walnut Grove
Bloomington, Indiana 47405
(812) 335-7636

National Cartographic Information Center
Entomology Bldg., Room 214
Purdue University
Lafayette, Indiana 47907
(317) 494-6305

United States Geological Survey
Western Distribution Branch
Federal Center
Box 25286
Denver, Colorado 80225
(303) 236-7477

All orders must be prepaid ($2.50/map)

A complete set of Topographic maps for the
State of Indiana is housed within the Department of Earth Sciences-Ag. At Vincennes
University and may be used for reference
by appointment.

MINERALS AND ROCKS IN V.U. KIT BY NUMBER

MINERALS

1. Galena
2. Sphaleite
3. Magnetite
4. Hematite
5. Bauxite
6. Pyrite
7. Calcite
8. Dolomite
9. Fluorite
10. Halite
11. Selenite Gypsum
12. Kaolinite
13. Olivine
14. Hornblende
15. Mica
16. Talc
17. Feldspar
18. Quartz
19. Garnet
20. Chrysotile Serpentine

ROCKS

21. Granite
22. Diorite
23. Syenite
24. Gabbro
25. Pumice
26. Scoria
27. Obsidian
28. Rhyolite
29. Basalt
30. Shale
31. Sandstone
32. Conglomerate
33. Limestone
34. Coal (Bituminous)
35. Slate
36. Phyllite
37. Schist
38. Gneiss
39. Quartzite
40. Marble

SES 100   Earth Science   Laboratory Exam I   "Sample Exam"
                                              (Exercises 1-3)

1. Latitude readings can exceed 90° but longitude
    readings cannot. A. True B. False

2. A degree of latitude is the same anywhere on earth;
    a degree of longitude changes from place to place
    depending upon the latitude. A. True   B. False

3. Correct the following latitude & longitude reading (if
    correct do nothing).
                         89° 56' 39" S.   112° 01' 40" E.

    _____   _____

4. What parallels are located at:

        23 1/2° South _____

        66 1/2° North _____

5. What is the longitude for Greenwich, England? _____

6. Latitude is to parallels what longitude is to _____

7. The only place where 2 lines of longitude are parallel is
    at the _____.

8. If 2 locations directly north & south of each other are
    45° of longitude apart, how many miles separte them?

9. The earth's axis is inclined _____° from the vertical
    toward the plane of the ecliptic.

10. Diagram the basic share of the analemma:

    What is the function of the analemma?

    _____
    _____

USE THE DIAGRAM PROVIDED TO ANSWER QUESTIONS 11-15:

11. What is the date? _____

12. Where are the vertical rays from the sun striking the
     earth? _____

13. What direction would people in Moose Jaw, Saskatchewan
     look to see sunrise? _____

14. How many hours of sunlight would there be at 71° S.
     latitude? _____

15. Will the length of the day be increasing or decreasing for people in North America over the next several months? _____

16. The earth is involved in at least 5 motions; name 4 of these movements: _____
_____.

17. Why did ancient people celebrate on their Winter Solstice? _____
_____
_____

18. The earth rotates on its axis at the rate of _____degrees/hour.

19. As one moves east across the earth's surface, hours of time are _____; if one crosses the International Dateline while moving in an easternly direction, a day must be _____.

20. How can one turn his alarm clock into a chronometer?
_____
_____.

21. It is impossible for 2 locations within the same hemisphere to be more than 12 hrs. apart in time.
A. True   B. False

22. Each day begins ont he East side (Western Hemisphere side) of the International Dateline and ends on the West side (Eastern Hemisphere side).   A. True B. False

23. Central Time Problem:
If it is 11:23 A.M. Monday at 83 degrees West, what is the time and day at 13 degrees East?

24. Sun Time Problem:
If it is 1:19 P.M. Wednesday at 123 degrees East, what is the "sun time" and day at 24 degrees East?

25. Chronometer Problem:
The chronometer rads 3:15 A.M. as you observe the noon sun-what is your location

SES 100   EARTH SCIENCE   LABORATORY EXAM II   "SAMPLE EXAM"
                                              (Exer. 4 - 7)

1. Metes and bounds surveying is more likely to be found in
    Vermont than Texas.   a) True   b) False

2. What is the general shape of the French Survey system
    lots? _____ Why? _____

    _____

3. Why are meridians resurveyed or off-set every 24 miles N
    or S of the base line in the "Congressional Township &
    Range System?

    _____

    _____

4. How many acres are in a quarter, quarter, quarter,
    quarter section? _____ acres

5. Diagram a section and place an X in the NE4, of the SW4,
    of the SE4, of the NW4, of the section.

6. There are ____ acres in a section of land, and ____ sq.
    miles in a Congressional Township.

7. What do the letters U.S.G.S. represent on a topographic
    map?

    _____

8. Datum plane for most maps is _____ _____.

9. What color are contour lines? _____

10. What is the symbol for a "bench mark" on a topographic
    map? _____

11. Which is more accurate: Aerial Photo Data, or Field
    Survey Data?   (Circle Answer)

12. On a map with a R.F. scale of 1:4000, a distance of 4"
    represents _____ miles.

13. Which is a smaller scale map?   1:50,000 or 1:5000?
    (Circle Answer)

14. Most topographic maps are drawn with _____ North at the top of the sheet.

15. Where can you purchase a topographic map of your home "stomping grounds?" _____

_____

16. Define the term: "relief" = _____

_____

17. What determines the contour interval for a map?

_____

18. A hachured contour line is between 2 contour lines of 800' and 900' value; it is closest to the 900' contour; the value of the hachured contour line is _____ feet.

19. Contour lines never _____ but can merge at _____.

20. Streams flow in the _____ direction of bending contour lines that cross them.

21. A map has a C.I. of 50'; the highest contour line is 2050', and the lowest contour line is 1000'. What is the maximum possible relief? _____ feet.

22. Most contour lines are smooth and flowing rather than sharp and irregular. a) true   b) false

23. The formula for Stream Gradient = _____

24. Give 2 rules for construction of a topographic profile:

   a. _____

   b. _____

25. As the relief of an area increases, the V.E. should

   _____.

LABORATORY EXAM III: Minerals

Student Name: _____   Lab. _____

==

Halite	Selenite Gypsum
Galena	Hematite
Dolomite	Hornblende
Mica	Feldspar
Bauxite	Magnetite
Garnet	Chrysotile Serpentine
Quartz	Fluorite
Calcite	Talc
Olivine	Kaolinite
Sphalerite	Pyrite

==

AT EACH STATION ANSWER QUESTIONS IN THIS ORDER:

    a.   Name a diagnostic trait;
    b.   Name the mineral;
    c.   Of what importance is the mineral?

==

KEEP EXAM SHEET COVERED!

BE SURE TO ANSWER QUESTIONS ON PROPER BLANKS!

STATION 1.

a. _____

b. _____

c. _____

STATION 2.

a. _____

b. _____

c. _____

STATION 3.

a. _____

b. _____

c. _____

STATION 4.

a. _____

b. _____

c. _____

STATION 5.

a. _____

b. _____

c. _____

STATION 6.

a. _____

b. _____

c. _____

STATION 7.

a. _____

b. _____

c. _____

STATION 8.

a. _____

b. _____

c. _____

STATION 9.

a. _____

b. _____

c. _____

STATION 10.

a. _____

b. _____

c. _____

STATION 11.

a. _____

b. _____

c. _____

STATION 12.

a. _____

b. _____

c. _____

STATION 13.

a. _____

b. _____

c. _____

STATION 14.

a. _____

b. _____

c. _____

STATION 15.

a. _____

b. _____

c. _____

STATION 16.

a. _____

b. _____

c. _____

STATION 17.

Name the mineral _____

STATION 18.

Name the mineral _____

LABORATORY EXAM IV: Rocks

Student Name: _____ Lab. _____

===

Pumice	Gabbro
Conglomerate	Sandstone
Gneiss	Basalt
Syenite	Schist
Quartzite	Coal (Bituminous)
Marble	Diorite
Obsidian	Slate
Shale	Rhyolite
Phyllite	Limestone
Granite	Scoria

===

AT EACH STATION ANSWER QUESTIONS IN THIS ORDER:

a. Is the rock Igneous, Sedimentary, or Metamorphic?
b. Is the rock intrusive or extrusive; clastic or non-clastic; foliated or non-foliated?
c. Name the rock.

===

KEEP EXAM SHEET COVERED!

BE SURE TO ANSWER QUESTIONS ON PROPER BLANKS!

STATION 1.

a. _____

b. _____

c. _____

STATION 2.

a. _____

b. _____

c. _____

STATION 3.

a. _____

b. _____

c. _____

STATION 4.

a. _____

b. _____

c. _____

STATION 5.

a. _____

b. _____

c. _____

STATION 6.

a. _____

b. _____

c. _____

STATION 7.

a. _____

b. _____

c. _____

STATION 8.

a. _____

b. _____

c. _____

STATION 9.

a. _____

b. _____

c. _____

STATION 10.

a. _____

b. _____

c. _____

STATION 11.

a. _____

b. _____

c. _____

STATION 12.

a. _____

b. _____

c. _____

STATION 13.

a. _____

b. _____

c. _____

STATION 14.

a. _____

b. _____

c. _____

STATION 15.

a. _____

b. _____

c. _____

STATION 16.

a. _____

b. _____

c. _____

STATION 17.

Name the rock _____

STATION 18.

Name the rock _____

I enjoyed the following laboratory topics the <u>most:</u>

1. _____

2. _____

3. _____

I enjoyed the following laboratory topics the <u>least:</u>

1. _____

2. _____

3. _____

Some very <u>good</u> aspects of the Earth Science lab are:

1. _____

2. _____

3. _____

Some aspects of the Earth Science Lab. that <u>need</u> <u>change</u> are:

1. _____

2. _____

3. _____

My laboratory instructor is _____; some of his

best teaching traits include _____

_____;

some areas for improvement include _____

_____

Overall, I would give the laboratory portion of Earth Science

the letter grade _____; I would give my laboratory instructor

the grade of _____.

LEST WE FORGET:

"THE LAST CHAPTER OF GENESIS"

In the beginning,
There was Earth, and it was with form and beauty.
And man dwelt upon the lands of the Earth, the meadows
and trees, and he said,
"Let us build our dwellings in this place of beauty."
And he build cities and covered the Earth with concrete
and steel. And the meadows were gone.
And man said, "it is good."

On the second day, man looked upon the waters of the
Earth and man said, "Let us put our wastes in the waters
that the dirt will be washed away."
And man did.
And the waters became polluted and foul in their smell.
And man said, "It is good."

On the third day, man looked upon the forest of the
Earth and saw they were beautiful.
And man said, "Let us cut the timber for our homes and
grind the wood for our use."
And men did. And the lands became barren and the trees
were gone.
And man said, "It is good."

On the fourth day man saw that animals were in abundance
and ran in the fields and played in the sun
And man said, "Let us cage these animals for our
amusement and kill them for our sport."
And man did. And there were no more animals on the face
of the Earth.
And man said, "It is good."

On the fifth day man breathed the air of the Earth.
And man said, "Let us dispose of our wastes into the air
for the winds shall blow them away."
And man did. And the air became heavy with dust and
choked and burned.
And man said, "It is good."

On the sixth day man saw himself, and seeing the many
languages and tongues, he feared and hated.
And man said, "Let us build great machines and destroy
these lest they destroy us."
And man built great machines and the Earth was fired
with the rage of great wars.
And man said, "It is good."

On the seventh day man rested from his labors and the
Earth was still; for man no longer dwelt upon the Earth.
And it was good.

Unknown